Springer Proceedings in Earth and Environmental Sciences

The series Springer Proceedings in Earth and Environmental Sciences publishes proceedings from scholarly meetings and workshops on all topics related to Environmental and Earth Sciences and related sciences. This series constitutes a comprehensive up-to-date source of reference on a field or subfield of relevance in Earth and Environmental Sciences. In addition to an overall evaluation of the interest, scientific quality, and timeliness of each proposal at the hands of the publisher, individual contributions are all refereed to the high quality standards of leading journals in the field. Thus, this series provides the research community with well-edited, authoritative reports on developments in the most exciting areas of environmental sciences, earth sciences and related fields.

Igor N. Vladimirov · Ming Jiang ·
Peter Ya. Baklanov
Editors

Resources, Environment and Regional Sustainable Development in Northeast Asia

Springer

Editors
Igor N. Vladimirov
Sochava Institute of Geography
Siberian Branch of the Russian Academy
of Sciences
Irkutsk, Russia

Ming Jiang
Northeast Institute of Geography
and Agroecology
Chinese Academy of Sciences
Changchun, China

Peter Ya. Baklanov
Pacific Geographical Institute
Far Eastern Branch of the Russian Academy
of Sciences
Vladivostok, Russia

ISSN 2524-342X ISSN 2524-3438 (electronic)
Springer Proceedings in Earth and Environmental Sciences
ISBN 978-3-031-28977-4 ISBN 978-3-031-28978-1 (eBook)
https://doi.org/10.1007/978-3-031-28978-1

This Springer imprint is published by the registered company Springer Nature Switzerland AG
The registered company address is: Gewerbestrasse 11, 6330 Cham, Switzerland

Contents

Assessment of the Relationship Between the Current Economic Development State and Environmental Pollution in the Republic of Buryatia, Russia

Alexey Bilgaev[1,2], Suocheng Dong[1], Erzhena Sadykova[2], Fujia Li[1,3](✉), and Cheng Hao[1]

[1] Institute of Geographic Sciences and Natural Resources Research, Chinese Academy of Sciences, Beijing 100101, China
lifj@igsnrr.ac.cn

[2] Baikal Institute of Nature Management, Siberian Branch of the Russian Academy of Sciences, 670047 Ulan-Ude, Russia

[3] Innovation Academy for Green Manufacture, Chinese Academy of Sciences, Beijing 100190, China

Abstract. The transition to green economic development is one of the priorities for the environmental-socio-economic development of the Baikal Natural Territory regions, including the Republic of Buryatia. Considering the special characteristics of the regional policy regarding Lake Baikal conservation, protecting the lake's ecosystem while maintaining sustainable economic development is one of the most important tasks at present. This study considers international experience in the green economic development transition, the relationship between economic growth rates and environmental pollution, and the decoupling of economic development and environmental pollution. In the example of the Republic of Buryatia, we assessed the relationship between the current state of economic development and environmental pollution using indicators of environmental intensity and the presence of the decoupling effect. The study's results made it possible to determine the current trend and prospects for the region's development. We gave recommendations on regional environmental policy improvement, as well as environmental measures in order to green production.

Keywords: Republic of Buryatia (Russia) · Lake Baikal · Green Economy · Decoupling

1 Introduction

Recently, the world has increased the urgency of the transition to green economic development – the Green Economy. The United Nations (UN) reports and documents on sustainable development note that forming a green economy is the basis for the transition to sustainable development [1, 2]. The United Nations Environment Program

I. N. Vladimirov et al. (Eds.): RERSDNA 2022, SPEES, pp. 1–8, 2023.
https://doi.org/10.1007/978-3-031-28978-1_1

(UNEP) defines the green economy as one that "improves human well-being and social equity while significantly reducing environmental risks and ecological scarcities" [2, 3]. In recent years, the green economy has emerged as one of several paradigms for fostering a growth path that integrates and reconciles economic, social, and environmental objectives [2, 4]. Key documents and reports of the UN conferences reflect green economy principles. Here we should highlight the fight against climate change (2015) and the transition to sustainable development based on the green economy formation (2012) [5]. Also, in 2015, the UN developed and adopted the sustainable development agenda until 2030, based on 17 sustainable development goals (SDG) [6].

The Baikal region is no exception. The transition to green economic development is one of the priorities of the ecological, social, and economic development of the Baikal Natural Territory (BNT) regions, including the Republic of Buryatia. In the current conditions, research areas for assessing the quality of environmental and socio-economic factors in the development of regions come to the fore. The economic activity of the regions and the assessment of the impact on the environment requires the use of special tools to consider the environmental component of the socio-economic development of territories. One of the key aspects of the concept of a "green" economy is the achievement of decoupling, which consists of a discrepancy or mismatch between economic growth rates, on the one hand, and resource consumption and negative environmental impact, on the other hand [7]. There are resource decoupling and impact decoupling. Resource decoupling implies a reduction in the use of primary resources (energy, water, minerals) per economic result unit. Impact decoupling is about increasing economic returns while reducing the negative environmental impact. Such impacts arise from resources extraction (pollution of groundwater from mining or agriculture), production (land degradation, waste, and emissions), the stage of use of goods (transport leading to emissions CO2), as well as at the post-consumption stage (waste and emissions) [8].

Lake Baikal has a special status, enshrined at the federal and world levels as a UNESCO World Natural Heritage Site (1996). The Baikal region, as part of the Russian part, includes the Republic of Buryatia (73%), Zabaykalsky Krai (21%), and Irkutsk Oblast (6%). The environmental regulation regime has been in force for more than 40 years and has a restrictive effect on the region's socio-economic development [9]. One of the main tasks is the scientific support of a balanced, sustainable development of the Republic of Buryatia, determined by a complex of environmental, economic, and social factors and conditions that form an integrated green sustainable development. The transformation of these factors towards integrating the green economy will increase the economic development of the Republic of Buryatia and the population's quality of life. In this paper, we evaluated the decoupling between environmental pollution and the economic development of the Republic of Buryatia. We made policy recommendations based on the results of the study. Protecting ecosystems while maintaining sustainable economic development in the region is one of the essential tasks at present.

2 Materials and Methods

In this study, the analysis of changes in the environmental sector of the Republic of Buryatia was carried out using the volumes of air, water, and waste pollutants per GRP unit, which show the relationship between economic development and the environment. In Russian practice, such an indicator is called "eco-intensity" [10]. A decrease in this indicator value – is a positive trend that shows the degree of technological shifts in the economy and characterizes the current environmental situation in the region [11].

We also calculated the decoupling coefficient D_t, which is used to identify the discrepancy between economic growth rates and environmental pollution. This coefficient is calculated by the formula [12]:

$$D_t = 1 - \frac{E_t/Y_t}{E_0/E_0}, \quad (1)$$

where E_0 and E_t are indicators characterizing the negative impact on the environment in the base and current periods; Y_0 and Y_t are indicators characterizing the economic result in the base and current periods, respectively.

The D_t coefficient shows the eco-intensity change compared to the beginning of the study period. Its negative value indicates the absence of the decoupling effect. On the other hand, the positive value of the D_t indicates multidirectional trends in economic development and anthropogenic impact: an increase in value added is accompanied by a decrease in the environmental burden on natural environments [7]. Initial data for the Republic of Buryatia were obtained from state reports on the environmental and socio-economic situation, statistical yearbooks of the Russian Federation, regions of Russia, and the Republic of Buryatia [13, 14].

3 Results and Discussion

The results of the analysis of the eco-intensity indicator of the Republic of Buryatia are influenced by the general dynamics of GRP growth and the dynamics of increasing air, water, and waste pollution. Figures 1, 2, and 3 show the dynamics of calculated indicators characterizing the level of environmental pollution: intensity of atmospheric air, wastewater, and waste for the analyzed period 2010–2019.

The analysis of air eco-intensity indicators showed that compared to the base year 2010, the indicator remained practically unchanged, and compared to the previous year, it decreased by 1.2 percent. In the period 2011–2015, there was a significant increase in emissions of pollutants into the atmospheric air, which affected the results. Since 2015, there has been a downward trend. Regarding wastewater eco-intensity, Buryatia showed a decrease of 28% by 2010 and 18% by the previous year – 2018. The analysis shows a downward trend for this indicator; in 2014–2016, there was a significant increase in wastewater discharges, with its peak in 2016. The eco-intensity of waste in the Republic of Buryatia in 2019 compared to 2010 increased by more than four times. At the same time, compared to the previous year, it decreased by 14%; according to this indicator, in previous years, there was an upward trend with a peak in 2018. A gradual decrease in the indicator of eco-intensity indicates a trend towards an environmental burden decrease with a relative increase in GRP, characterized as a positive process. However, the rate remains relatively very high, which applies to air and water pollution. This indicates a low level of technical equipment in the enterprises of the Republic.

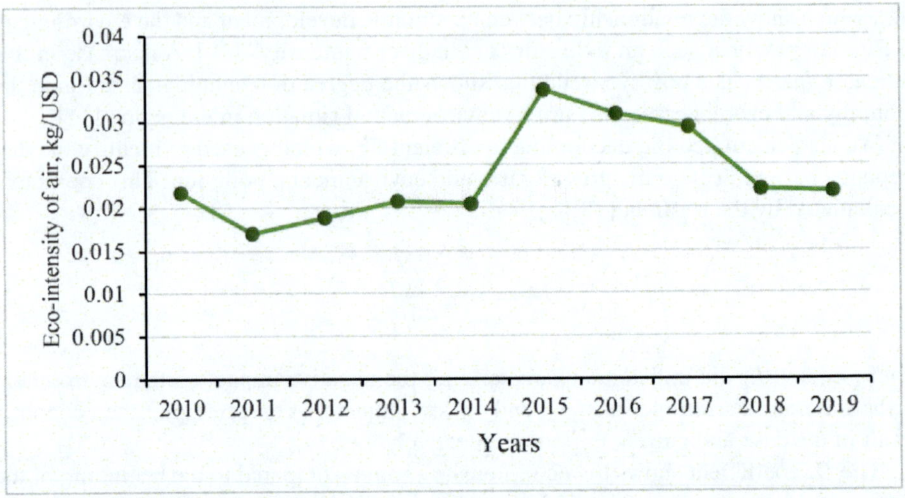

Fig. 1. Eco-intensity of air in the Republic of Buryatia for 2010–2019.

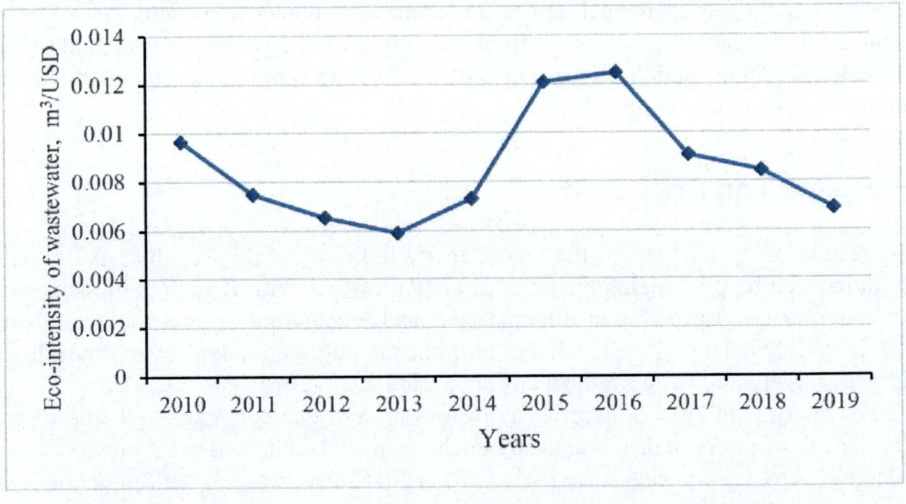

Fig. 2. Eco-intensity of wastewater in the Republic of Buryatia for 2010–2019.

To identify the effect of decoupling on the development of the economy of the Republic of Buryatia, we calculated the decoupling coefficient D_t for the period 2010–2019 for three indicators - emissions of pollutants into the atmosphere, discharges of polluted wastewater, solid waste (Table 1). The most unfavorable situation has developed in solid waste, the value of the studied indicator ($D_t = -3.31$). The main waste generators in the Republic of Buryatia are the extraction of ores and sands of precious metals, rare metals, coal mining, brown coal, cement, lime, and plaster. The main reason for the unsatisfactory environmental state of the mining sector is that the funds allocated by the organizations are not enough to overcome the negative trends that have developed in the subsoil use system, especially when preserving a unique reservoir – Lake Baikal. Regarding pollutant emissions, we observe a slight negative change in the study period ($D_t = -0.005$). Thus, we observe a favorable situation only for discharges of polluted wastewater ($D_t = 0.29$). However, this may be due to the low quality of monitoring and not an actual decrease in the level of negative impact.

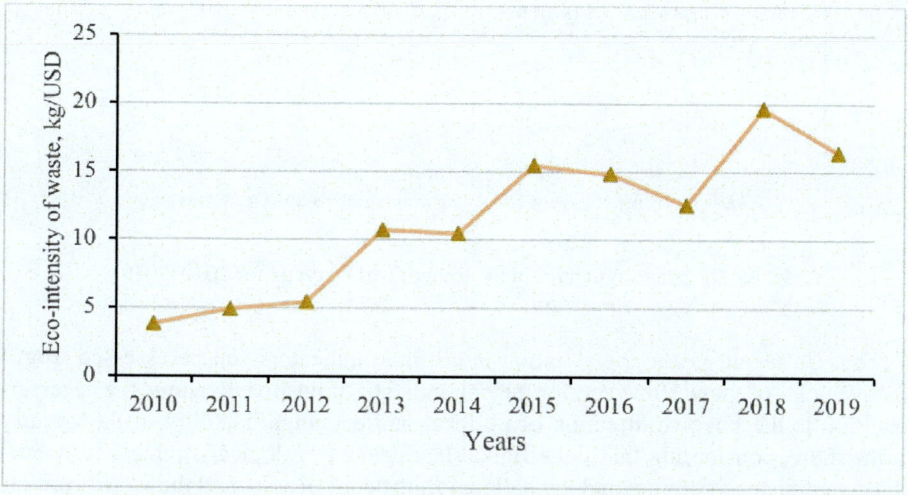

Fig. 3. Eco-intensity of waste in the Republic of Buryatia for 2010–2019.

Table 1. D_t values in the Republic of Buryatia for 2010–2019.

	D_t of air	D_t of wastewater	D_t of waste
2011 to 2010	0.217	0.225	−0.295
2012 to 2011	−0.108	0.127	−0.109
2013 to 2012	−0.101	0.097	−0.958
2014 to 2013	0.015	−0.238	0.023
2015 to 2014	−0.645	−0.653	−0.477
2016 to 2015	0.082	−0.032	0.043
2017 to 2016	0.049	0.271	0.153
2018 to 2017	0.246	0.071	−0.564
2019 to 2018	0.013	0.182	0.163
2019 to 2010	– 0.005	0.29	−3.31

Figure 4 shows the dynamics of the decoupling coefficient D_t for 2010–2019 in the Republic of Buryatia.

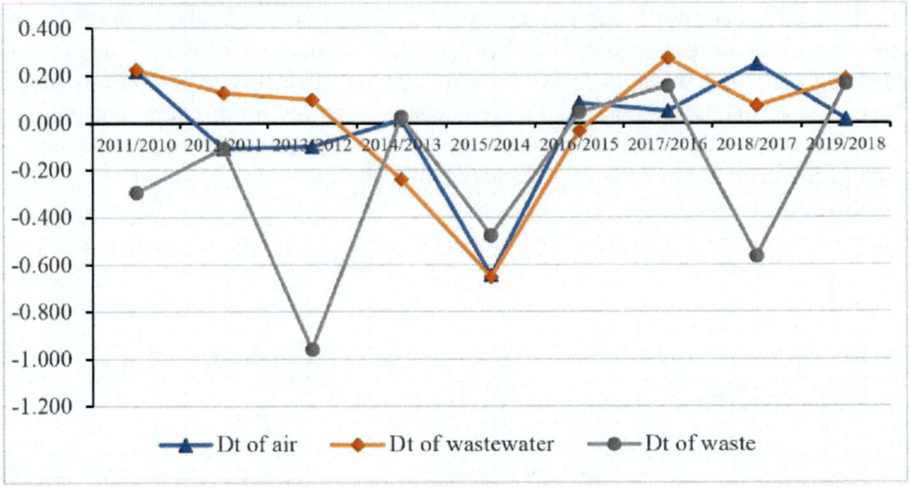

Fig. 4. D_t values dynamics in the Republic of Buryatia for 2010–2019.

Thus, in recent years, the situation in all three indicators can be assessed from a positive point of view. However, for the continuation of positive dynamics, a necessary condition is the positive direction of political and economic stability in the country. Furthermore, considering that Lake Baikal is a "global ecological strategic resource", which significance goes beyond the national framework, Russia and the world community should be interested in the green development of the Baikal region, including the Republic of Buryatia. Implementing strategic directions for countries with economies transitioning to green economic development is possible through the region's innovation, optimal resource use, and scientific and educational potential to achieve sustainability of the Republic of Buryatia industries. Furthermore, achieving the optimal reproduction of material goods and services is possible using the advantage that the Baikal region is a model territory that meets the primary imperatives of sustainable development. This requires conditions for business development, state support for investment projects, activation of institutional transformation processes, overcoming poverty, and making attractive living conditions [11].

China is an example that, during the last decade, has made great efforts to reduce air pollution by implementing laws, regulations, and environmental policies and implementing emission reduction projects for atmospheric pollutants such as sulfur dioxide (SO_2) and nitrogen oxides (NO_x), and soot. Since 2012, China has made a strategic decision to promote the construction of an "ecological civilization", putting forward the concept that "clear waters and green mountains are as good as mountains of gold and silver" [15]. In 2013, China implemented the Action Plan for the Prevention and Control of Air Pollution (the Action Plan) [16]. In 2018, China promulgated the "Three-Year Action Plan to Beat Air Pollution" (the Three-Year Action Plan) – an updated version of the Action Plan. However, by 2019 57% of China's cities had not met air quality standards yet [17]. The long-term goal is to build a "Beautiful China" that achieves sustainable development and a highly developed ecological civilization by 2035.

4 Conclusion

The study showed that in recent years there had been a positive trend in all three indicators of eco-intensity. However, to continue the positive dynamics, a necessary condition is the positive direction of political and economic stability in the country. The obtained quantitative estimates of the indicators of eco-intensity and the decoupling coefficient can be applied in assessing the socio-economic development towards the transition to a "green" economy for the BNT regions, particularly the Republic of Buryatia. Furthermore, the study's results can help solve the problems of predicting the development of environmental consequences of technogenic pollution of regions and taking appropriate preventive management measures to prevent them.

Acknowledgements. This research was funded by the Class A strategic leading science and technology project of Chinese Academy of Sciences (Grant number XDA20030203), Key Collaborative Research Program of the Alliance of International Science Organizations (Grant number ANSO-CR-KP-2020-02, ANSO-CR-KP-2020-06), National Natural Science Foundation of China (Grant number 32161143029, 42071282) and State assignment of BINM SB RAS (Grant number AAAA-A21-121011590039-6).

References

1. Bobylev, S., Perelet, R.: Sustainable development and the "green" economy in Russia: current situation, problems and prospects. In: Bobylev, S., Perelet, R., Davydova, A., Kokoreva, A.: Sustainable development in Russia, pp. 11–19. Russian-German Environmental Information Bureau, Saint Petersburg, Russia (2013)
2. Bilgaev, A., Dong, S., Li, F., Cheng, H., Sadykova, E., Mikheeva, A.: Assessment of the current eco-socio-economic situation of the Baikal Region (Russia) from the perspective of the green economy development. Sustainability **12**, 3767 (2020)
3. UNEP: Towards a Green Economy: Pathways to Sustainable Development and Poverty Eradication. Nairobi. https://sustainabledevelopment.un.org/content/documents/126GER_synthesis_en.pdf. Accessed 25 Jun 2022
4. Halsnæs, K., et al.: Climate change mitigation policy paradigms – national objectives and alignments. Mitig. Adapt. Strateg. Glob. Chang. **19**, 45–71 (2014)
5. Bobylev, S., Goryacheva, A.: Stable development: purposes and investments. Soc. Power **5**(67), 61–64 (2017)
6. United Nations: The 2030 Agenda for Sustainable Development, New York. http://www.un.org/ga/search/view_doc.asp?symbol=A/RES/70/1&Lang=E. Accessed 20 Jun 2022
7. Zabeline, I.: Decoupling in environmental and economic development of regions-participants of crossborder cooperation. Econ. Soc. Changes Facts Trends Forecast. **12**(1), 241–255 (2019). https://doi.org/10.15838/esc.2019.1.61.15
8. Decoupling Natural Resource Use and Environmental Impacts from Economic Growth. UNEP (2011). http://www.gci.org.uk/Documents/Decoupling_Report_English.pdf. Accessed 17 Jun 2022
9. Ecological atlas of the Baikal basin. V.B. Sochava Institute of Geography SB RAS, Irkutsk (2015)
10. Saktoev, V., Sadykova, E.: Sustainable Development of Regional Economic System with Environmental Regulation. Ekonomika, Moscow (2011)

11. Bilgaev, A., Dong, S., Li, F., Cheng, H., Tulohonov, A., Sadykova, E., Mikheeva, A.: Baikal region (Russia) development prospects based on the green economy principles. Sustainability **13**(1), 157 (2021)
12. Naqvi, A., Zwickl, K.: Fifty shades of green: Revisiting decoupling by economic sectors and air pollutants. Ecol. Econ. **133**, 111–126 (2017). https://doi.org/10.1016/j.ecolecon.2016.09.017
13. Regions of Russia: Socio-Economic Indicators. Rosstat, Moscow, Russia (2021)
14. Federal State Statistic Service Republic of Buryatia Branch. https://burstat.gks.ru/publication_bur. Accessed 1 Jun 2022
15. Li, J., Hou, L., Wang, L., Tang, L.: Decoupling analysis between economic growth and air pollution in key regions of air pollution control in China. Sustainability **13**(12), 6600 (2021). https://doi.org/10.3390/su13126600
16. Feng, Y.Y., Ning, M., Lei, Y., Sun, Y.M., Liu, W., Wang, J.N.: Defending blue sky in China: effectiveness of the "Air pollution prevention and control action plan" on air quality improvements from 2013 to 2017. J. Environ. Manag., 252 (2019)
17. Ministry of Ecology and Environment. Report on the State of the Ecology and Environment in China 2019. http://english.mee.gov.cn/Resources/Reports/. Accessed 18 Jun 2022

New Guidelines and Risks for Asian Russia in the Implementation of the "One Belt – One Road" Project Under Current Geopolitical and Global Climate Changes

Tamir A. Boldanov[1,2,3](✉), S. Dong S[1], Fujia Li[1], Hao Cheng[1], and Tsogto Zh. Bazarzhapov[1,2,3]

[1] Institute of Geographic Sciences and Natural Resources Research, Chinese Academy of Sciences, Beijing, China
tboldanov2018@igsnrr.ac.cn
[2] University of Chinese Academy of Sciences, Beijing, China
[3] Baikal Institute of Nature Management, Siberian Branch, Russian Academy of Sciences, Ulan-Ude, Russia

Abstract. The life of modern society of Asian Russia is governed by the political processes occurring on the western borders of Russia as well as by global warming which together create new vectors of economic development throughout the entire territory of North Asia. With the imposition of Western sanctions, however, there is taking place a reorientation of the export-import relations of Russia with a focus on the development of contacts with countries of the Asia-Pacific region. In these conditions, the program of development of international transport corridors entitled "One Belt – One Road" is gaining new momentum; it has a direct bearing on the prospects of the economic growth of Russia, North China, Mongolia and other countries located in the zone of influence of historical civilizational roads, such as the "Great Silk Road" and "Great Tea Road". This article considers the present situation from the perspective of both economic and physical geography as well as the possible problems associated with eastward reorientation of the main economic flows and the possible ways out of this situation. Economic factors are considered, such as energy flows and commodity-money relations, as well as ecological factors: climatic changes, the influence on agriculture and tourist flows that have shaped themselves in the modern realities.

Keywords: "One Belt – One Road" · Economic corridor · Transport corridor

1 Introduction

One of the priorities of the development of the Russian state for the 21st century involves the economic development of its eastern regions. However, in modern conditions of complicated and strained geopolitical situations, a steady growth of innumerable sanctions that are more harmful to the side that imposes these sanctions and at the same time

I. N. Vladimirov et al. (Eds.): RERSDNA 2022, SPEES, pp. 9–15, 2023.
https://doi.org/10.1007/978-3-031-28978-1_2

the effectiveness of the Russian economy in conditions of the openness of the world economy is increasingly determined by the entry into the system of international trade ties and, largely, into the system of transcontinental transport corridors. Furthermore, it must be borne in mind that all over the world there are taking place events associated with the development of green energy, global climate change, preconditions for a new industrial revolution, and a reduction in the carbon economy [1].

In recent years, the Government of RF has been actively considering the document on comprehensive development of the territory of Asian Russia on the basis of the transport spatial-logistic Unified Eurasia corridors built upon creating two spatial transport-logistic corridors between Europe and Asia with reliance on the high-speed railroad complex and the Northern Sea Route [2]. The authors of the project propose to interconnect the economic interests of 23 Russian regions, create several millions of jobs, and to ensure a growth of the country's GDP to 4–5%. Its cost is estimated at 18 trillion rubles, and its payback time at eight years.

According to most analysts, the existing transport networks are overloaded and cannot cope with the requirements of natural resource exports. In this regard, there arise questions as to how and with what effectiveness the capabilities of railroads are used, what do they transport, who benefits from it, whether is it possible to improve the effectiveness of these transport operations, and others. Analysis of operating results of the Russian Railways [3] shows that in recent years most of transport operations correspond to coal (372 million tons), oil and oil products (232 million tons), and iron and manganese ores (120 million tons). Up to 10 thousands of coal cars are loaded every day in Kuzbass alone. With an increase in quality and quantity of high-speed roads, a part of cargo transportation, especially to short distances, will inevitably be shifted to motor transport.

Most experts believe that in the near future the consumption of hydrocarbons will alter the energy market. According to the Ministry of Energy, the world will cross the peak of oil consumption within 10–15 years, which was necessarily followed by a drop associated with an increase in the output of electric vehicles [4]. However, according to data of the International Energy Agency, the transition to "green" energy has already decreased the world's consumption of coal by 7% against the year 2019, whereas the demand for renewable energy sources increased by 0.9%. On the whole, the proportion of coal in the world's energy consumption reached 27%, such were the minimum indicators for 16 years 16 [5]. It should be noted that for the last five years the prices of power-generating coal varied from $ 42 to 95 per ton, which corresponds to the product cost and zero profitability for Russian coal companies, even without transportation costs which almost double the export expenses and are unlikely to survive without budgetary subsidies.

2 Object and Materials

In the first place, it should be noted that One Belt – One Road" is the shortest terrestrial transport route connecting European countries with China. Unfortunately, its today's condition does not correspond to the growing cargo and transport operations and requires its radical reconstruction with the laying of a second track and its electrification, with the possibility of operative communication of cars from the narrow Chinese standard to the Russian broad-gauge track [6].

Nevertheless, Russia is implementing plans for the expansion of the traffic capacity of the main transport routes: Transsib and BAM, to 180 million tons. In 2020, 37 million tons of coal were delivered to China, which is 1 million tons larger than one year earlier. India is planning to increase the purchases by 6–7%. However, the situation in the world coal market is under heavy pressure of “green” energy. Thus, in China, the solution of ecological problems is aimed at restricting carbon emissions, and the attainment of the emission peak is planned for the year 2030, and the scheduled year of carbon neutralization is 2060 [7]. For the last two years the consumption of gas in China has increased by 33% to reach 40% of the world’s consumption in 2024. In modern conditions, however, plans are changed relatively rapidly, and it is always necessary to look at the prospects and calculate the moves based on the incomplete picture of data.

Given the energy deficit in the economy of Mongolia and North China, there is a need to create the East-Asian energy ring with the possibility of supplying electric energy from the cascade of the Angara HEPs and the Zeya and Bureya HEPs which do not have an adequate number of Russian consumers. The supplies of Russian electricity to the industrial centers of Mongolia can remove from the agenda the construction of hydroelectric power plants in the Mongolian part of the Selenga river basin which creates the ecological threat to the ecosystem of Lake Baikal as its main tributary [8].

The construction of the Power of Siberia 2 gas pipeline has a strategic significance for Russia, Mongolia and China which implies supplies with the most environmentally friendly type of fuel from the deposits of Yamal to Eastern Siberia and further to our southern neighbors. Thus there arises a real opportunity to substitute the refusal of European gas consumers by supplies to China and Mongolia.

3 Results and Discussion

It is fundamentally important to note that the new gas pipeline route will be built along the industrial centers of the Yenisei Siberia, the Angara region and Cisbaikala which will be given an additional incentive for the development of an ecologically safe economy. In addition, there arises a real opportunity to additionally provide gas supplies to the central ecological zone of Lake Baikal and reduce air emissions from the numerous settlements along the lake coast [9].

At the suggestion of the Limnological Institute and the Baikal Institute of Nature Management, the gas pipeline can be placed along the shortest route on the bottom of Lake Baikal along the Transsiberian Railroad to Ulan-de and further toward Mongolia. In this case, it will be possible to extend the pipeline toward Transbaikalia which is also among the energy-deficient subjects of the Far Eastern Federal District.

Furthermore, the observations that were made during the expedition of the deep-water manned “Mirs on Baikal” vehicles suggest that on the bottom of the lake there are numerous outcrops of hydrocarbons in gaseous, liquid and solid state and they do not affect negatively the lake ecosystem [10]. Moreover, habitats of protozoa emerge nearby oil outcrops, which constitute the beginning of the trophic chain for subsequent phagues.

On the other hand, the project of construction of the Power of Siberia 2 gas pipeline provides for the variant of its laying via the Tunka valley which crosses Tunkinsky National Park. In accordance with Russian environmental legislation, any kinds of economic activity, including the construction of pipelines, are prohibited in specially protected areas of such a rank. No such restrictions exist for the central ecological zone, and only some alterations to the decree of the government [11].

The Western sanctions and restrictions on travel to traditional tourism centers of Europe and the Mediterranean have drastically altered the tourist flows, whereas the flow of guests to Baikal has increased dramatically. There emerged the system of charter tourism, and winter tourism began to enjoy special popularity. Tourists from tropical countries are eager to see Baikal's transparent ice. "Baikal Fishing" attracts more and more participants, and the same applies for "Baikal Mile" among motorists.

At a meeting of the Council on Science and Education V.V. Putin drew particular attention to the necessity of developing scientific and educational tourism [12]. From this perspective, the Baikal region is the home for the unique astrophysical observatories of the Russian Academy of Sciences carrying out observations of solar activity and near-terrestrial and deep space. The world's largest underwater neutrino telescope (Baikal Gigaton Volume Detector, Baikal-GVD) is in operation in South Baikal with the purpose of recording the most elusive elementary particles—neutrinos.

The touristic interests of our countries are united by the location of the route of the Great Tea Road [13]. The Buddhist Traditional Sangha Center of Russia, Ivolginsky Datsan, is located in Buryatia, where the imperishable body of the famous Khambo Lama Dashi-Dorzho Itigilov resides, attracting pilgrims from around the world. On the boundary with Mongolia, in the city of Kyakhta, on the occasion of the 175th anniversary of the Russian Geographical Society, the Memorial Arch "Gates to Asia" was built, with the images of the great travelers and explorers of Central Asia headed by N.M. Przhevalsky etched in bronze.

Buryatia is one of the few world centers which can offer medical tourism based on using traditional prescriptions of oriental medicine, and medications reproducing the experience of Tibetan lamas, including their pulse diagnosis [14].

Global climate changes decrease the productivity of agricultural land of many traditional centers of agriculture in Asia and in Africa. The increasing occurrence rate of tropical hurricanes and droughts increase the threat of famine for many developing countries. On the other hand, Transbaikalia and Mongolia that are located in the heart of the Asian continent less frequently experience climatic disasters. The last natural landscapes—traditional centers of nomadic pastoralism of the peoples of the Great Steppe are also located there. In Mongolia alone, with its population of less than four million, the number of domestic livestock reaches 50–60 million head [15]. Furthermore, these animals mostly use natural pastures with a special taste of meat products that are very popular in the world market as organic products without the use of chemical and artificial preparations.

Diversification of the food industry in the neighboring country makes it possible to create food products oriented to special religious groups of population, including halal and kosher products. Mongolia is already delivering, as prototype products, halal food to Iran which is subjected to the most comprehensive control. An identical promising market is represented by Muslim countries of Southeast Asia.

In Buryatia, the federal law "On the protection of Lake Baikal" has been in force for many years now; it has by-laws prohibiting the use of chemicals in agriculture. Legislation thereby supports the manufacture of organic food, acquiring an increasing demand in the market of high quality goods [11].

At the same time it should be noted that political and climatic changes involve certain risks for the economic development of Asian Russia. Primarily, because of the transit restrictions on the western borders of Russia, Asian partners have to look for ways to bypass the Russian territory which is loosing its transit potential [16]. For West China the TRACECA international transport corridor is becoming more attractive; international freight is transported through Kazakhstan and Central Asia on the Caspian Sea and further through Georgia and Turkey to Southern Europe [17].

With global melting of the Arctic ice, the transit potential of the Northern Sea Route increases. In this case, however, international cargoes move still further from the Russian ports pole ward and directly, with the shortest possible distance, are transported from the Chkchi Sea to Norway. Moreover, with an increase in freight turnover of the Northern Sea Route, there is a decrease in the volume of transit container freight in Russian Far Eastern ports, and in transportation volumes via Transsib.

4 Conclusions

By and large these factors predetermine the new "sounds" for the economic interests of Russia, Mongolia and China adjoining the existing Ulan-Ude – Ulaanbaatar – Beijing railroad.

The main risk for the development of Asian Russia remains the reduction of its population, primarily highly skilled personnel. The years of economic reforms saw a complete loss of economic and moral incentives of the attraction of youth to new industrial facilities. It is necessary not only to re-establish the mechanisms of providing production facilities with staff but also to create new ones. In addition to increased pay and northern benefits, it is necessary to decrease for the local population payments for housing, social infrastructure, gasoline and fuels, i. e. the advantages which can only be used in places of permanent evidence. To stop the outflow of the population, it is necessary to introduce increased pay for the work experience.

Thus the military and economic conflict with Western countries in connection with the events in the Ukraine alters radically and for long and, maybe, for ever all the stereotypes of international relations between Russia and the West which were established after the disintegration of the Soviet Union.

It is becoming increasingly clear that the new Russian economy should be focused on a cluster development of Asian Russia. It must, to some degree, adopt the best practices from the period of the first Stakin's five-tear plans. When in anticipation of a great war the head of the state, surrounded by political enemies, by a volitional method, in the complete absence of highly skilled personnel, transport infrastructure, and financial and material resources, in the shortest possible time, constructed aircraft plants in Novosibirsk, Irkutsk, Ulan-Ude, Komsomolsk-on-Amur and in Arsenyev, the Norilsk Industrial Complex beyond the Arctic Circle, Kuzbas and Turksib, and started to construct BAM.

Today, many top executives have found plenty of reasons for delaying the economic growth of the Russian periphery. However, there is no longer time for discussion and debate. There is an urgent need for quick solutions and implementation of the "One Belt – One Road" program, and cooperation with our eastern neighbors offers one of the few ways out of the current crisis situation.

Acknowledgments. This research was done with the financial support within the CAS-TWAS President's PhD Fellowship Program for international scholars. Key Collaborative Research Program of the Alliance of International Science Organizations (Grant number ANSO-CR-KP-2020-02, ANSO-CR-KP-2020-06). Class A strategic leading science and technology project of Chinese Academy of Sciences (Grant number XDA20030203), National Natural Science Foundation of China (Grant number 32161143029, 42071282).

References

1. Dong, S., et al.: Opportunities for the development of North Asia in the "One Belt – One Road" project: problems and solutions. In: Proceedings of the Environmental Science and Technology International Conference (ESTIC 2021), Netherlands (2021)
2. Comprehensive development of the territory of the Russian Federation on the basis of transport spatial-logistic corridors. In: Kozlov, V.V., Makosko, A.A. (eds.) Current Problems of Implementation of the Megaproject "United Eurasia: TEBR–IETN". Nauka, Moscow (2021)
3. Annual Report of JSC RRW for 2020: Analysis of operations results (2020). https://ar2020.rzd.ru/pdf/ar/ru/performance-overview_analysis-operating-results_transportation-logistics.pdf. Accessed 15 Jul 2022
4. Seleznev, M.: On the needle: the oil curse will be removed from Russia within 10–15 years. Profil 14–15, 52–54 (2021). https://profile.ru/economy/na-igle-neftyanoe-proklyatie-snimut-s-rossii-cherez-10-15-let-844667/. Accessed 15 Jul 2022
5. Katkov, M., Volobuev, A.: A gloomy future is predicted for coal miners. Vedomosti **29**, 13–14 (2021). https://www.vedomosti.ru/business/articles/2021/01/29/855869-rossiiskim-ugolschikam. Accessed 15 Jul 2022
6. Tulokhonov, A.K., Boldanov, T.A.: Political geography of Russian border areas in conditions of globalization: theory, challenges, solutions. Region Ekonomika i Sotsiologiya **1**, 203–235 (2021). https://doi.org/10.15372/REG20210108
7. Zhang, J., Chung, Y.: On the way to reduce emissions. Kitai **184**, 36–37 (2021)
8. The construction of the Erdeneburen hydroelectric power plant will begin in April // Mongolia segodnya. https://www.mongolnow.com/stroitelstvo-gidroelektrostantsii-erdeneburen-nachnetsya-v-aprele/. Accessed 15 Jul 2022
9. Official Website of the Gazprom Company: Gazprom instructed to proceed to pre-investment phase in Power of Siberia 2 project. https://www.gazprom.ru/press/news/2020/march/article502469/. Accessed 15 Jul 2022
10. Tulokhonov, A.K.: The Mirs of Baikal's depths: results and reflections. Ekos, Ulan-Ude (2010)
11. Federal law of 1 May 1999 no. 94-FZ: On the protection of Lake Baikal (1999). https://base.garant.ru/2157025/. Accessed 15 Jul 2022
12. Official network resource of the President of Russia. List of instructions on the basis of results of a joint meeting of the State Council and the Council on Science and Education. http://kremlin.ru/acts/assignments/orders/67752. Accessed 15 Jul 2022
13. Mamirkulova, G., Mi, J., Abbas, J., Mahmood, S., Mubeen, R., Ziapour, A.: New Silk Road infrastructure opportunities in developing tourism environment for residents better quality of life. Glob. Ecol. Conserv. **24**, e1194 (2020)

14. Solpina, N.G., Ivanova, A.S.: The development of scientific and educational tourism in the Baikal region on the absis of astronomic-geodetic objects. Izvestiya irkutskogo gosudartsvennogo universiteta. Seriya: Nauki Zemle **2**(1), 160–172 (2009)
15. Mongolian statistic information service. https://1212.mn/BookLibraryDownload.ashx?url=Mongolian%20Statistical%20Yearbook%202021.pdf&ln=Mn. Accessed 15 Jul 2022
16. Vinokurov, E., Lobyrev, V., Tikhomirov, A., Tsukarev, T.: Silk road transport corridors: assessment of trans-EAEU freight traffic growth potential (May 11, 2018). Report 49. EDB Centre for Integration Studies, Saint-Petersburg (2018). https://ssrn.com/abstract=4094996. Accessed 15 Jul 2022
17. Tulokhonov, A.K: The geographical space of the new Russia: on the past. Present and Future. Ekos, Ulan-Ude (2020)

Placer Gold Mining in the Zabaikalskii Krai and Its Role in the Development of Municipal Districts

Irina P. Glazyrina[1,2](✉) and Irina S. Kalgina[2]

[1] Institute of Natural Resources, Ecology and Cryology, SB RAS, Chita, Russia
i_glazyrina@bk.ru

[2] Transbaikal State University, Chita, Russia

Abstract. The article deals with studying the impact of placer gold mining on socio-economic processes in the east of Russia on the example of the Zabaikalskii Krai. The quantitative assessment of the development of the Zabaikalskii Krai found that there are short-term benefits from the placer gold mining during the mining season in the municipalities of excavating sites. However, there is no reason to say that it has a positive impact on socio-economic and demographic processes over the medium term. The author arrived at the conclusion, that the role of placer gold mining in improving the quality of life of the local population and demographic stabilization in the transboundary regions is great and revealed the need to carry out a number of institutional changes. They are: a radical redistribution of income within the budget system in this segment of subsurface management; an increase in payment rates within the limits of maximum permissible, temporarily agreed and excess discharges to aquatic biological resources for damage, since the existing payment system has almost completely lost its regulatory functions and is now frankly imitative; a change in the procedures for commercial access to resources so that the local community can really participate and play a significant role in decision-making.

Keywords: Placer gold · Gold mining · Local budgets · Demographic dynamics

1 Introduction

This article is devoted to the role of placer gold mining in the development of the eastern territories of Russia on the example of the Zabaikalskii Krai. The discussion about the prospects and expediency of this segment of subsurface use is relevant in the context of the growing negative impact on natural systems. We can talk about the growing polarization of experts' views and approaches to these problems. Industry experts often consider environmentalists' concerns about the environmental impact degree of placer gold mining to be overstated and unjustified. However, in the data stream in various information resources, there is more and more evidence that this type of economic activity is accompanied by the destruction of natural systems, generates serious environmental

I. N. Vladimirov et al. (Eds.): RERSDNA 2022, SPEES, pp. 16–25, 2023.
https://doi.org/10.1007/978-3-031-28978-1_3

conflicts of mining companies with the population of adjacent territories, and at the same time does not meet the expectations of economic benefits for the local community.

Placer gold mining in the Zabaikalskii Krai is a significant segment of subsurface use and a significant source of budget revenues. At the same time, it is one of the most environmentally hazardous types of nature management, accompanied not only by spot pollution, but also by the destruction of valuable river ecosystems and the loss of their ecological functions. Therefore, the balance of economic and environmental interests in the placer gold mining remains one of the most relevant in the field of natural resource management.

The article presents the results of empirical studies of these processes from the point of view of the concept of “green economy” and “green growth”, which include not only environmental indicators, but also issues of the quality of life of the local population and demographic dynamics.

2 Placer Gold Mining in the Zabaikalskii Krai

In the Zabaikalskii Krai, the extraction of ore gold has been growing rapidly in recent years, and one of the reasons for this was the launch of production at the newly built Bystrinskii Mining and Processing Plant [1, 2]. Nevertheless, the growth of placer gold production also continues (Table 1) and the number of licenses issued is growing (see Fig. 1).

Table 1. Gold mining in the Zabaikalskii Krai for 2012–2021 [4–6].

Source of gold	2012	2013	2014	2015	2016	2017	2018	2019	2020	2021
Ore, t	6,1	5,3	5,9	7,9	7,7	12,2	14,6	18,8	25,4	29,5
Placer, t	5,7	6,4	6,4	6,9	7,9	9,8	10,2	11,1	9,1	11,2
Total, t	11,9	11,7	12,3	14,8	15,6	22,0	24,8	29,9	34,5	40,7

It is necessary to note the cardinal change in the structure of the license “package” issued annually in the Zabaikalskii Krai [3]. Between 2010 and 2015, the issuance of licenses in the category “exploration survey and assessment” was one-time occurrences. More licenses were issued in the category of “geological survey, exploration and mining”. The vast majority of licenses issued were in the category of “geological survey, exploration and mining” and their share was steadily exceeding 80%. Starting in 2016, the share of exploration survey and assessment licenses that do not involve resource extraction began to increase sharply. Their number has increased hundreds of times over the past four years, and their share had exceeded 75% by 2021 (see Fig. 2).

The extraction of placer gold in Siberia and the Far East is accompanied by increasing damage to river ecosystems. There are more and more reports about this in the media and social networks. Public environmental organizations are engaged in systematic monitoring of these processes. Integrated data is posted, among other things, on the website

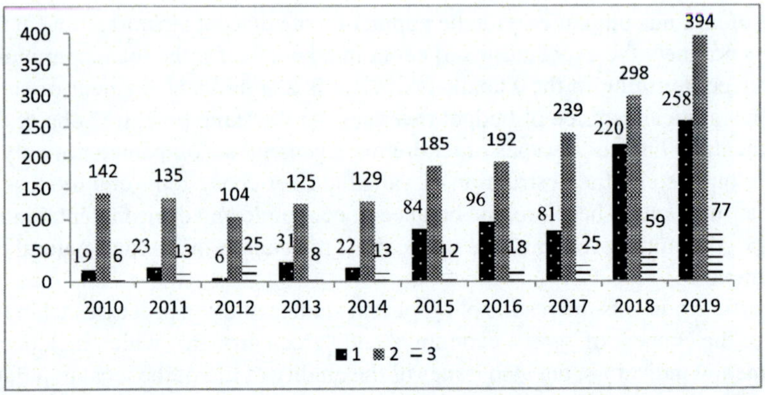

Fig. 1. Dynamics of issuance of licenses for exploration and production of placer gold, number of licenses/year [7, 8], authors' calculations. 1 – Amur region, 2 – Zabaikalskii Krai, 3 – Khabarovsk Krai.

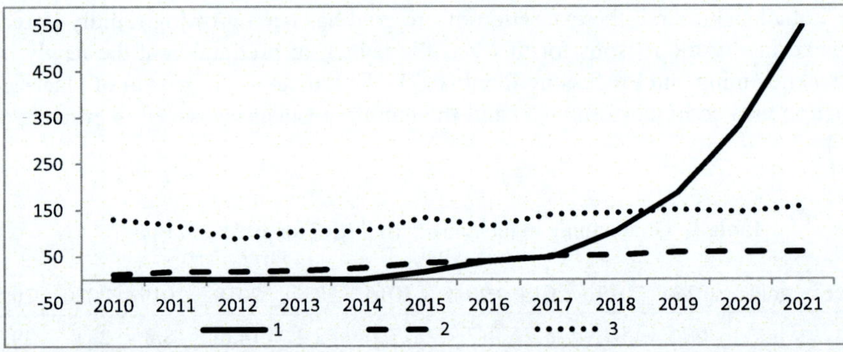

Fig. 2. Dynamics of issuing licenses for placer gold mining in the Zabaikalskii Krai, 2010–2021 [7], calculations of the authors. 1 – exploration survey and assessment, 2 – geological survey, exploration and mining, 3 – exploration and mining.

of the Russian division of the World Wildlife Fund. The main problem is the river pollution and the destruction of floodplain ecosystems. The effects include a decrease in water quality (including drinking water) and deterioration of conditions for agriculture, reduction of fish productivity and deterioration of habitat conditions of aquatic animals, damage to tourist and recreational resources, loss of the potential of territories to provide ecosystem services [9–11], etc. Ultimately, this leads to a deterioration in the quality of life and rural exodus, which traditionally depends on the natural capital of the territories, and to an aggravation of social tension. [12]. The scale of the negative impact is indicated by the detected pollution of more than one and a half thousand kilometers of rivers in the Far East according to space monitoring data, which in 2020 was carried out with the support of WWF Russia.

In almost all gold-mining regions of the Far East, there is an increase in the number of companies receiving quotas for the extraction of relatively small volumes of gold (up to

100 kg, in some cases less than 20 kg). In the Zabaikalskii Krai, 46 gold mining companies received such quotas in 2022, which is more than half of their total number. Experts say that mining cannot be profitable at such volumes and with the implementation of the necessary environmental measures and the creation of an appropriate environmental protection infrastructure. The fact that in these cases the project documentation is approved by the authorized state bodies indicates the low quality of their work.

The analysis of the effectiveness of environmental institutions in coordinating environmental and economic interests in the placer gold mining sector was carried out in [13] using regulatory impact assessment matrices, the methodology of which dates back to the works of Elinor Ostrom [14]. An analysis of payments for damage to aquatic ecosystems was made using the example of companies that extracted about a quarter of placer gold in the Zabaikalskii Krai in 2016. It is shown that in this economic sector, the mechanism of charges for using natural resources has almost completely lost its regulatory functions and is now already imitative in nature. The rates of payments, including for emergency discharges into water bodies, is so insignificant (several hundred rubles a year with the turnover of tens, sometimes hundreds of millions) that does not create motivation for environmental protection activities even within the framework of project documents agreed by state bodies. It is pointless to rely on the environmental awareness of the companies themselves to reduce costs in these conditions. Hence, even minimal revenues from subsoil users in settlement budgets are significant for them in a context of chronic under-funding, as shown in the next section. Therefore, local administrations are interested in the development of placers on their territory, and often turn a blind eye to environmental violations. This study also shows that the expansion of the scope of state environmental control due to subsoil users will not lead to an unacceptable increase in their transaction costs [15] and will not become a critical factor for the profitability of the sector.

When analyzing the current practice, experts come to the conclusion that due to the small number and outdated technical equipment, environmental authorities are not able to carry out proper environmental control over a large number of subsoil users in vast territories. Economic regulatory mechanisms do not work due to the insignificance of the rates (even for excess emissions). And, most importantly, the issuance of licenses for the use of subsurface areas is legally arranged in such a way that not only the local community, but also regional authorities and municipalities are practically excluded from decision-making procedures: «This leads to the issuance of licenses for gold mining to sites that are important as habitats for rare and economically valuable species of flora and fauna, as well as those that play an important role in the economic activity of the local population as places of traditional residence of indigenous peoples of the North and the Far East, hunting grounds, locations of apiaries, hayfields and pastures. The destruction of these lands and the loss of valuable natural resources during the development of deposits have a negative impact on other, no less significant sectors of the local population» [12]. Proposals to change the existing practice were formulated by several environmental organizations with the participation of the scientific community and sent to the Government of the Russian Federation in 2020, but the further fate of the document is unknown [8].

3 Gold Mining and Budget Revenues

Proponents of placer gold mining often appeal to its positive impact on socio-economic processes: opportunities for small and medium-sized businesses to participate, create additional jobs, increase revenues to local budgets, etc. In this section, we have tried to assess whether those hopes are justified, and whether the full potential is actually being realized.

The importance of the district budgets' own revenues is often underestimated by the governing bodies, since the policy of "equalization" still compensates for their insufficiency to some extent. Often those with lower incomes simply receive more transfers from regional budgets. However, own funds and funds from other levels of the budget system have different "quality". The availability of sufficient, stable and independent own incomes allows municipalities to plan the development of territories at least in the medium term, to set and solve the most urgent tasks that are often simply "not visible" from higher levels. In addition, it makes the local authorities more responsible to the citizens, at whose expense these budgets are replenished, contribute to the growth of "inclusiveness" of socio-economic relations in general [16, 17].

It is possible to assess revenues to budgets of all levels from the extraction of placer gold in the Zabaikalskii Krai. Since there is not enough information in the public domain to make accurate calculations (including the fact that it is protected by a trade secret), this article makes an assessment based on the scheme from [18], in which a forecast was made assuming that the total volume of placer gold production would be 8.7 tons. However, in fact, it amounted to 11.3 tons, and appropriate adjustments have been made in this work. The results are given in Table 2.

Table 2. Estimated tax revenues from placer gold mining in the Zabaikalskii Krai for 2019 [4, 7, 18, 19], authors' calculations.

Levels of the budget system	Tax revenues, million rubles	Share of total tax payments, %
Federal budget, Including MET (40%), Including fuel excise (100%) Including income tax (3%)	1986 299 1041 82	63,34
Regional budget, Including MET (60%); Including income tax (17%), Including personal income tax (85%)	1112 449 467 194	35,54
Local budgets, Including personal income tax (15%)	39 34	1,12
Total	3137	100

The main tax revenues from this segment of subsurface use are directed to the federal and regional budgets, and almost two-thirds of all revenues in the first one. This seems unfair to the population of territories that experience negative consequences of this activity, even in cases when companies carry out environmental protection measures in accordance with approved project documentation. In many cases, residents note a significant deterioration of roads and bridges (the quality of which is often not the best anyway) mining equipment, pollution of water sources, reduction of hayfields and damage to fisheries.

Figure 3 shows information on the average per capita income of district budgets in the Zabaikalskii Krai in nominal terms.

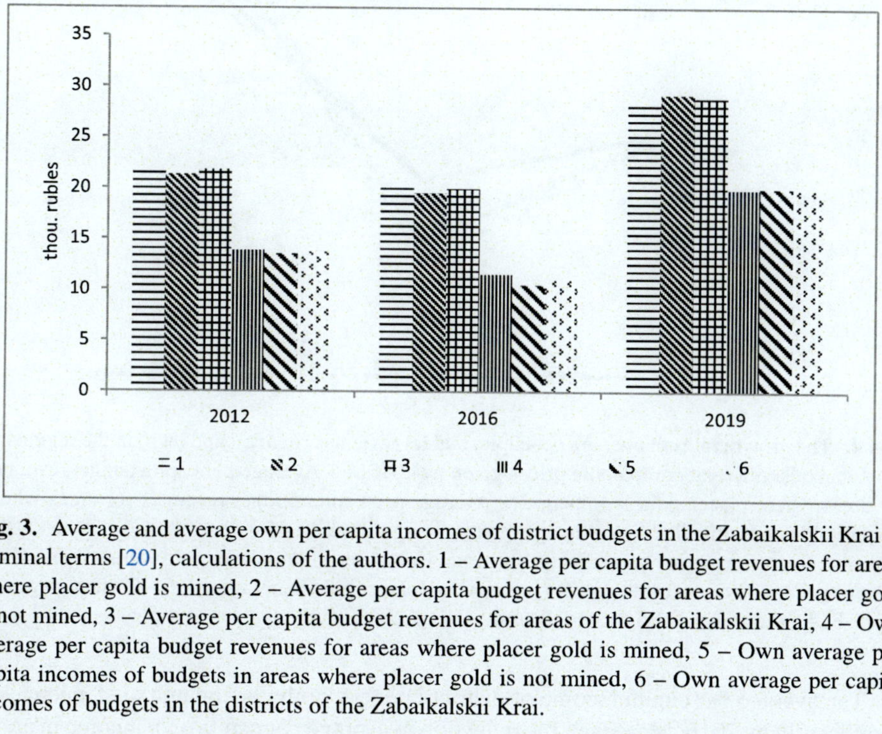

Fig. 3. Average and average own per capita incomes of district budgets in the Zabaikalskii Krai in nominal terms [20], calculations of the authors. 1 – Average per capita budget revenues for areas where placer gold is mined, 2 – Average per capita budget revenues for areas where placer gold is not mined, 3 – Average per capita budget revenues for areas of the Zabaikalskii Krai, 4 – Own average per capita budget revenues for areas where placer gold is mined, 5 – Own average per capita incomes of budgets in areas where placer gold is not mined, 6 – Own average per capita incomes of budgets in the districts of the Zabaikalskii Krai.

The total per capita incomes of district budgets are not expected to differ in areas where placer gold is mined and where placer gold is not mined – the policy of budget alignment is aimed at this. At first glance, it is surprising that there is practically no difference between own income depending on gold mining. But this does not indicate the presence/absence of income itself, but the structure of the tax system in Russia. The main tax source is wages, and, judging by the data in Table 2 and Fig. 3, in annual terms in the districts, regardless of whether gold is mined there or not, it differs little from the income of "state employees" who provide the bulk of personal income tax revenues in the districts. All income containing the natural resource rent itself goes to higher budgets.

District budget revenues in nominal terms, as a matter of fact, are growing in all districts. However, if we consider these dynamics in comparable prices, the picture changes somewhat. Figure 4 shows it as a percentage in relation to 2012, comparable prices are also shown for 2012.

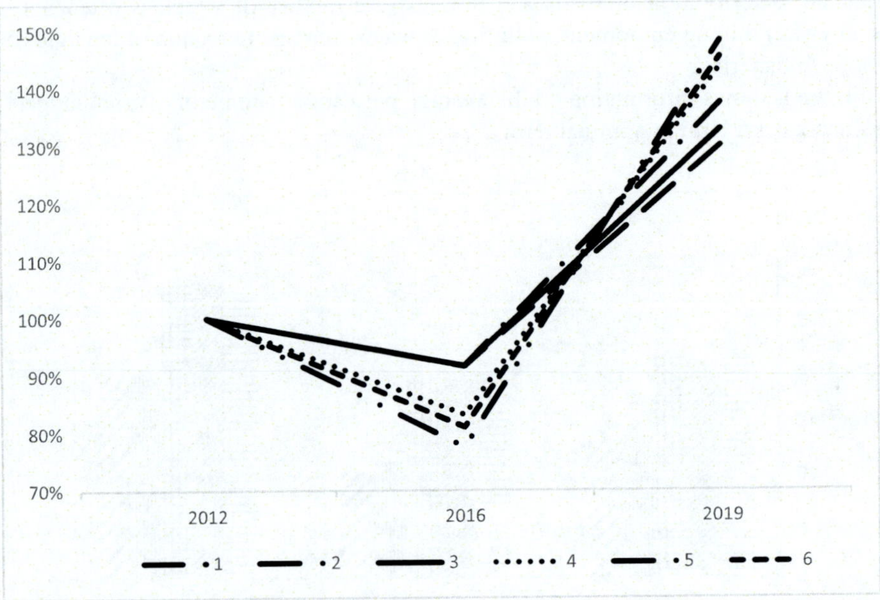

Fig. 4. The growth rates of average per capita budget revenues (general and own) in the regions of the Zabaikalskii Kraiat comparable prices given by 2012. 1 – Average per capita budget revenues for areas where placer gold is mined, 2 – Average per capita budget revenues for areas where placer gold is not mined, 3 – Average per capita budget revenues for districts of the Zabaikalskii Krai, 4 – Own average per capita budget revenues for areas where placer gold is mined, 5 – Own average per capita budget revenues for areas where placer gold is not mined, 6 – Own average per capita budget revenues for the districts of the Zabaikalskii Krai.

The average per capita income of district budgets, both general and own, decreased significantly by 2016. Moreover, the reduction in average per capita own income in areas where placer gold is not mined was more significant (−22%) than where it is (−17%). This can be explained by the fact that the wage fund in the former is more dependent on the budget, and after the 2014 crisis, transfers from the regional budget decreased. But by 2019, all these indicators had grown and even exceeded the level of 2012 (see Fig. 4). At the same time, the average per capita incomes of district budgets grew, both general (37%) and own (48%), where placer gold is not mined. The growth rates of average per capita incomes of district budgets where placer gold is mined were slightly lower (30% and 45%, respectively).

4 Population Trends

Placer gold mining is carried out, as a rule, in areas with a predominantly rural population, which experiences negative environmental consequences both in everyday life and in economic activities. On the other hand, they have the right to count on economic benefits from the use of resources on their territory. The overall "balance" of these multidirectional consequences is largely expressed in demographic trends. Table 3 presents some data and calculations characterizing these processes.

Table 3. Population dynamics in the districts of the Zabaikalskii Krai [7, 20], calculations of the authors.

Data on districts	Population, thou people		Changes over 10 years, %	Number of licenses issued over 10 years
	01.01.2010	01.01.2020		
Districts, where placer gold is mined, except for the Chita (rural) district	373241	334623	– 10,4	1943
Districts, where placer gold is not mined	294756	263163	– 10,7	-
All districts of the Zabaikalskii Krai, except for Chita and Chita (rural) district	667997	597786	– 10,5	1943
Nerchinsk-Zavodskii district	10767	8986	– 16,5	232
Baleiskii district	20416	17222	– 15,6	167

The population has been steadily declining, and generally has decreased by more than 10% over 10 years in all districts of the Zabaikalskii Krai, with the exception of Chita and the Chita district adjacent to the regional center. The rate of population decline in the areas of placer gold mining practically does not differ from the rates in other areas. And in the Nerchinsk-Zavodskii and Baleiskii districts, which account for more than 20% of all licenses issued over 10 years, the reduction was more than 1.5 times higher than the regional average (Table 3). That clearly shows that the long-term practice of placer gold mining cannot be considered a factor of demographic stability in the transboundary regions, as many supporters of its expansion claim.

5 Conclusion

The studies of Russian and foreign scientists devoted to the development of countries with rich natural resource potential have revealed the importance of the transition of governing institutions from extractive to inclusive [16], etc. *Extractive* institutions include institutions focused on extracting economic or political rent and redistributing it in the interests of elites – small groups with economic and/or political power. They are opposed to *inclusive* institutions that provide favorable conditions for productive activity and increase its effectiveness in the interests of the majority of citizens. The studies of Russian scientists, including those led by Academician P.A. Minakir [21, 22] indicate that the extractivity of Russian institutions under the Far Eastern transformation does not decrease. On the example of the placer gold mining sector, the "extractivity" of decision-making procedures on access to resources is particularly evident. Another important factor hindering "inclusivity" is the current format of fiscal federalism, in which about 1% of tax revenues fall into district budgets, and rental income completely goes to higher budgets.

The quantitative assessments of the development of the districts of the Zabaikalskii Krai show that there are short-term benefits from the development of placer gold sites during the mining season in those municipalities where it is conducted. In the medium term, however, there is no reason to assert that this type of activity has a positive impact on socio-economic and demographic processes. According to the level of individual and social well-being, gold mining areas differ insignificantly from those where gold mining is not carried out, in particular, this concerns budget security per capita. The growth rate of own budget revenues in areas where placer gold is mined is even slightly lower than in those where gold is not mined.

Population decline in the areas of placer deposits development is no less intense than in other areas. This is one of the main problems of the border territories of the east of the Russian Federation.

Nevertheless, in the current geopolitical conditions, it is hardly advisable to completely abandon the extraction of placer gold. This resource has a certain export potential in the Asian markets, while other export opportunities have been largely lost. However, it is necessary to carry out a number of institutional changes to make this a real factor in improving the quality of life of the local population and demographic stabilization in the border regions [23, 24]. Firstly, it is necessary to radically change the distribution of income within the budget system in this segment of subsoil use. Secondly, the rates of payments have to be increased within the limits of maximum permissible, temporarily agreed and excess discharges and for damage to aquatic biological resources. The size of the current payments does not contribute not only to the prevention of emergency discharges, but also to the introduction of more gentle extraction technologies. It is urgently necessary to change the payment system for the negative impact on water bodies as a whole, which has almost completely lost its regulatory functions and is now frankly imitative. And, most importantly, to change the system of commercial access to resources so that the local community can really participate in decision-making and play a significant role in this.

Acknowledgement. The work was supported by grant No. 350-GR of the Council for Scientific and Innovative Activities of the Transbaikal State University. The research methodology was developed within the framework of the state task of the IPEC SB RAS under project No. FUFR-2021-001 "Mechanisms for ensuring economic sustainability and environmental safety in a new model of development of the regions of the East of the Russian Federation in terms of cross-border relations and global challenges of the 21st century".

References

1. GRKB Homepage. https://www.grkb.ru/. Accessed 15 Jun 2022
2. Simonova, E.: Golden Rivers. Orange, Vladivostok (2012)
3. Latysheva, M.: Current state, prospects and problems of the gold mining industry of the Zabaikalskii Krai. In: IOP Conference Series: Earth and Environmental Science. 2nd «EBT 2021», pp. 1–6. Publishing, Chita (2021)
4. Information on the state and prospects for the use of the mineral resource base of the Zabaikalskii Krai as of March 15 (2021). https://www.rosnedra.gov.ru/data/Fast/Files/202104/a1fe101c2b86b515977f10aa9c3f192b.pdf. Accessed 15 Jun 2022
5. Gold mining in the Zabaikalskii Krai. https://zolotodb.ru/article/11232. Accessed 13 May 2022
6. State balance sheet for gold. Zabaikalskii Krai. Report of Rosnedra (explanatory note as of 01.01.2020)
7. ROSNEDRA Homepage. https://www.rosnedra.gov.ru. Accessed 25 May 2022
8. Our position, https://zolotari.net/position, last accessed 2022/05/25
9. Titova, G.: Assessment of ecosystem services: the potential for application in practice. Vestn. Zabaik. Univ. **3**(118), 179–191 (2015)
10. Farley, J.: Ecosystem services: the economic debate. Ecosyst. Serv. **1**, 40–49 (2012)
11. Mikheev, I.: Ecosystem services as a tool for protecting the biocenoses of small rivers. Fish Farming Fish. **4**, 3–7 (2015)
12. WWF Russia. https://wwf.ru/resources/news/amur/wwf-rossii-vynosit-na-obsuzhdenie-proekt-pozitsii-prirodookhrannykh-organizatsiy-po-rossypnoy-dobych/?fbclid=IwAR03IHVdMI_%20FlFcJsLEmLTCZm72cngMnMpfIEWFoIXlS5W9NKz3I6mI7S4g. Accessed 25 May 2022
13. Glazyrina, I., Mikheev, I., Eloyan, A.: Accommodation of ecological and economic interests in placer gold mining. Geogr. Nat. Resour. **38**(3), 275–280 (2017)
14. Ostrom, E., Burger, G., Field, C., Norgaard, R., Policansky, D.: Revisiting the commons: local lessons, global challenges. Science **284**, 278–282 (1999)
15. Marshall, G.: Transaction costs, collective action and adaptation in managing complex systems. Ecol. Econ. **88**, 185–194 (2018)
16. Acemoglu, D., Robinson, J.: Why Nations Fail. The Origins of Power, Prosperity, and Poverty, 529 p. Crown Business Publishing, New York (2012)
17. Natkhov, T., Polishchuk, L.: The political economy of institutions: the importance of being inclusive. J. New Econ. Assoc. **2**, 12–38 (2017)
18. Gold mining in Transbaikalia. Magazine «Gold mining». https://zolotodb.ru/article/11230. Accessed 13 May 2022
19. NALOG Homepage. https://www.nalog.gov.ru. Accessed 13 May 2022
20. ROSSTAT Homepage. https://rosstat.gov.ru. Accessed 13 May 2022
21. Minakir, P.: Far Eastern institutional innovations: imitation of a new stage. Spat. Econ. **15**(1), 7–17 (2019). https://doi.org/10.14530/se.2019.1.007-017
22. Minakir, P., Naiden, S.N.: Social dynamics in the Far East: a defect in ideas or a failure of institutions? Reg. Econ. Sociol. **3**(107), 30–61 (2020)
23. Kryukov, V., Kolomak, E.: Spatial development of Russia: main problems and approaches to overcoming them. Sci. Works Free Econ. Soc. Russ. **227**(1), 92–114 (2021)
24. Kuleshova, V. (ed.): The Mineral Resource Sector of Asian Russia: How to Ensure Social and Economic Returns? 352 p. Publishing House of the Institute of Economics and Organization of Industrial Production of the Siberian Branch of the Russian Academy of Sciences, Novosibirsk (2015)

Ecological and Economic Assessment of the Use of Wood Fuel for Heat Supply Purposes

Elena Gubiy(✉) and Elena Maysyuk

Melentiev Energy Systems Institute, Siberian Branch of the Russian Academy of Sciences, 130 Lermontov Street, Irkutsk 664033, Russia

egubiy@ise.irk.ru

Abstract. Enterprises of the timber industry lose a large amount of wood in the form of waste, which is advisable to use for energy purposes. A comprehensive ecological and economic assessment of the efficiency of its use is necessary to justify the introduction of wood fuel. A method of ecological and economic assessment of the cost of reduced emissions for energy facilities is proposed. It is based on the ratio of the costs of implementing environmental technology and the environmental effects of its implementation. The proposed method makes it possible to estimate the environmental effect, namely, the amount of reduced or prevented emissions of pollutants and greenhouse gases into the atmosphere in value terms. The method was tested on the example of replacing coal with wood fuel (fuel pellets and wood chips) in the boiler houses of the coastal areas of Lake Baikal. The study showed that wood fuel is environmentally attractive and economically promising. The efficiency of its use depends on a number of factors: the prices of coal and wood fuel, the remoteness of the boiler house from the place of fuel production, the transport infrastructure of the area, the power of the heat source. A significant environmental effect was revealed: a reduction in pollutant emissions by 94%, a reduction in greenhouse gas emissions into the atmosphere by 99.9% and a reduction in the formation of ash and slag waste by 92–91%.

Keywords: Biofuels · Pollutant missions · Ash and slag waste · The cost of reduced emissions · Greenhouse gas emissions

1 Introduction

Timber industry enterprises (TIE) lose a large amount of wood in the form of waste, which is advisable to use for energy purposes in the form of fuel [1]. Between 2000 and 2018, global emissions of carbon dioxide (CO_2) alone increased by almost 50% [2], with the largest share of emissions coming from the energy sector. Therefore, the measures provided for by the Paris Agreement [3] for the transition to more environmentally friendly energy resources and technologies are certainly justified.

The implementation of any environmental protection measure requires comparative ecological and economic assessments. This study proposes to use an ecological and economic method to assess the ratio of the costs for implementing a specific event (e.g.,

I. N. Vladimirov et al. (Eds.): RERSDNA 2022, SPEES, pp. 26–35, 2023.
https://doi.org/10.1007/978-3-031-28978-1_4

replacing coal with wood fuel) and the environmental effect obtained from it. The method was tested on the example of replacing coal with wood fuel in boiler houses of coastal areas of Lake Baikal.

2 Objects and Methods

2.1 Ecological and Economic Assessment

This paper proposes to use an indicator hereinafter referred to as the cost of reduced emissions for the purposes of ecological and economic assessment. This indicator is a criterion that reflects the environmental component of environmental protection measures and enables the assessment of the level of costs for the production of the final product. The cost of reduced emissions is calculated for each ith energy facility according to the formulas (1–2):

$$CRE = (C_i - C_0)/(M_0 - M_i), \tag{1}$$

$$CRE \rightarrow \min, \tag{2}$$

where C_i is the unit costs of the energy facility after the introduction of environmental technology i, \$/Gcal; C_0 is the unit operating costs before the introduction of environmental technologies, \$/Gcal; M_0 is the specific emissions of pollutants before the introduction of environmental technologies, t/Gcal; M_i is the specific emissions of pollutants after the introduction of environmental technology i, t/Gcal.

Different environmental measures, including the transition to other environmentally friendly fuels, have different structures and the amount of energy production costs. The difference in unit costs is interpreted as the amount of increase in unit operating costs for environmental protection measures aimed at reduction of pollutant emissions.

The proposed indicator is a close analogue of the now popular "Levelized Cost of Carbon Abatement" (LCCA) [4–7]. However, there are a number of differences. In the general case of LCCA, when assessing the level of costs aimed at minimizing environmental damage, the numerator (1) uses an indicator reflecting the discounted amount of capital and annual costs. The proposed methodological approach uses the annual current costs of energy production. Firstly, this approach allows for a more objective comparison of changes in the current tariff for the consumer in situations where the existing equipment is completely worn out. Secondly, capital investments are often allocated in the form of targeted one-time payments that do not require payback and are aimed at achieving a significant socio-economic effect or a specific environmental goal in order to reduce the anthropogenic impact in areas of special nature use. An example of the use of such funds is the implementation of the Federal Target Program "Protection of Lake Baikal and socio-economic development of the Baikal Natural Territory for 2012–2020".

The suggested approach evaluates the economic effect of environmental measures, namely the amount of reduced (or prevented) emissions in value terms. Furthermore, additional environmental effects can be obtained: an assessment of the reduction of greenhouse gas emissions (for energy facilities – CO_2) and the amount of reduction in the mass of ash and slag waste (ASW).

Methods and approaches to environmental assessment are regulated based on available background information. Due to the difficulty of obtaining reliable information about direct measurements at energy facilities, it is possible to estimate the mass of emissions by calculation based on officially approved methods [8–10].

To quantify greenhouse gas emissions from energy enterprises in Russia, Order No. 300 of 30.06.2015 [11] is in effect. The guidelines and manuals of this Order have been developed in accordance with the methodology of the Intergovernmental Panel on Climate Change (IPCC) [12]. This is especially true for the use of wood fuel, when the world community has accepted the fact that there are no CO_2 emissions during its combustion. If CO_2 regulation is adopted using a carbon tax, the cost of which is currently being discussed in the range from $40 to $120/ton, accounting for CO_2 emissions will play a positive stimulating role in the adoption of alternative measures.

In addition, the study assessed CO_2 emissions not only from the combustion of fuels, but from the production and transportation of these fuels to the place of use. In the energy facilities under consideration, coal is used, which is extracted at regional coal mining enterprises in an open manner. As the existing estimates show, there is no carbon dioxide emission during open-pit coal mining. However, the IPCC guidelines [12] mention the presence of CO_2 in coalbeds and adjacent interstices. To estimate CO_2 emissions from open-pit coal mining, the IPCC recommended emission factor of 1.5 m^3 per ton of coal produced.

A special calculator "GHG calculator version 1.2" was used to calculate greenhouse gas emissions from biofuel production [13]. This tool is designed to perform calculations of greenhouse gas emissions during the life cycle of wood pellets with the quality mark "EN plus".

To estimate the mass of ASW from energy facilities in this study, we used the method of material and raw-materials balance, where the amount of ASW is determined as the sum of the mass of slags formed during the combustion of solid fuel in a boiler house and the mass of fly ash caught from exhaust gases in special trapping devices.

2.2 Description of the Research Object

The presented method of ecological and economic assessment has been tested on the example of a number of boiler houses in the coastal areas of Lake Baikal, which cover three districts of the Irkutsk Oblast and partially or completely five districts of the Republic of Buryatia. This territory has a special status of nature management, which is defined by the Federal Law of the Russian Federation "On the protection of Lake Baikal" of 01.05.1999 [14] and is recognized by UNESCO as an example of an outstanding freshwater ecosystem [15]. According to [14], economic activities that have a negative impact on the unique ecological system are prohibited in the central ecological zone of the Baikal Natural Territory. In order to protect it, economic activity and the existing energy infrastructure should be reoriented to eco-friendly ways of management.

One of the possible ways to reduce the load on the natural environment may be the replacing to eco-friendly fuels. This is especially true for low-power coal-fired boilers, in which there is currently no flue gas cleaning or it is not carried out properly.

According to the research of the Melentiev Energy Systems Institute of SB RAS [16], there are 104 municipal boiler houses of various capacities on the territory of the coastal areas of Lake Baikal. These heat sources use almost all types of energy resources as energy carriers, however, the share of coal in total fuel consumption is 97.7%.

According to the authors' estimates, the emissions from the energy facilities amount to 17–24 thou tons/year; SAW – 40–60 thou tons/year; CO_2 – up to 520 thou tons/year depending on the air temperature and the duration of the heating period.

The analysis is based on the following initial data on prices: fuel pellets – 97.2 $/t; fuel chips – 20.8 $/t; transport of wood fuel by road – 0.22 $/km•m^3, by rail – 0.04 $/km•m^3; coals by brand, $/t: Cheremkhovskii – 30.5 (transport by rail only) and 37.4 (transport by rail and road), Cheremkhovskii DKOM – 55.5, Haranutskii – 58.3, Tugnuyskii – 62.5 and Borodinskii – 49.3.

3 Results and Discussion

3.1 Transport Factor

Wood fuel is distinguished by a relatively low energy density. Therefore, its transport over long distances significantly increases the final cost of energy. The permissible distance limits depend on the energy density of a particular type of wood fuel and the transport infrastructure of the area [17, 18].

The influence of the transportation distance of wood fuel on the cost of thermal energy has been studied. Figure 1 shows how the fuel component of pellets and wood chips changes in the cost of thermal energy depending on the distance over which cargo transportation is carried out on the example of road transport. When transported for 50 km, the fuel components in the cost of 1 Gcal of thermal energy obtained by burning both types of biofuels are equal. When transporting at a distance of less than 50 km, it is advisable to use fuel chips, and at a distance of more than 50 km – pellets.

3.2 Carbon Dioxide Emissions from Fuel Production and Transport

The calculation of CO_2 emissions from the production (extraction) and transport of coal and wood fuel has shown that open-pit coal mining and its transport is much more environmentally friendly compared to the production of wood fuel. Specific CO_2 emissions from the extraction of one ton of coal are more than 50 times lower than this indicator for the production of one ton of pellets and 10 times lower for the production of wood chips. Table 1 shows the specific CO_2 emissions from open-pit coal mining and from the production of wood fuel and from their transport.

3.3 Ecological and Economic Assessment of the Replacement of Coal with Wood Fuel

Boiler houses located no more than 50 km from the enterprises supplying wood chips are accepted as potentially suitable for the use of fuel chips. There are 4 such boiler houses identified in the coastal areas of Lake Baikal. All of them provide thermal energy to social objects, the annual volume of heat production at each of them is less than 1 thou Gcal/year.

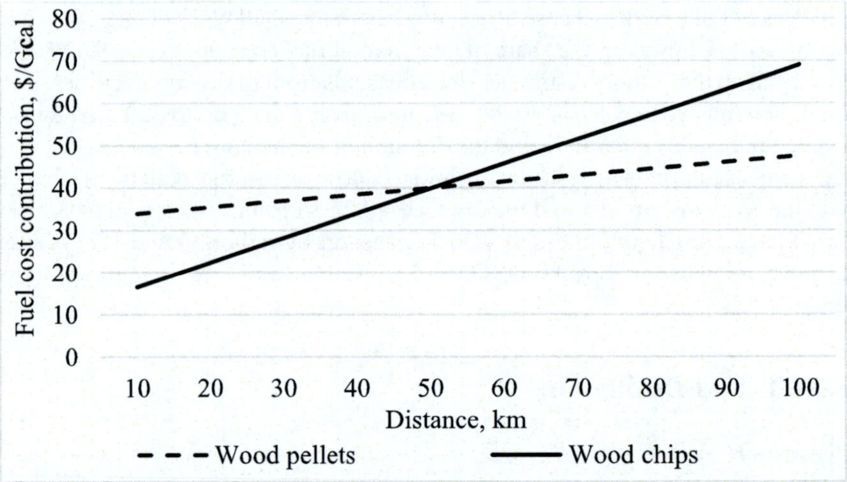

Fig. 1. Changes in the fuel components of wood pellets and wood chips in the cost of thermal energy depending on the range of cargo transportation.

Table 1. Specific CO_2 emissions from fuel production, t CO_2/t.

Fuel	Production			Transport (100 km)	Total
	Electrical energy	Thermal energy	Wrapping		
Coal	0.002			0.008	0.002
Pellets	0.08	0.0005	0.03		0.108
Chips	0.03	0.0002	0.00		0.030

All boilers with an installed capacity of less than 10 Gcal/h are considered as potentially suitable for the use of fuel pellets, except: boilers that currently use firewood (8), electric boilers (25) and boilers suitable for the use of fuel chips (4). Thus, 54 boiler houses were selected that are potentially suitable for replacing coal with pellets.

To assess the economic efficiency of replacing coal with wood fuel, a comparative analysis of the specific current costs for the production of thermal energy at existing coal-fired and wood-fired boilers was carried out. All boiler houses are divided into three groups according to boiler capacity. It is obvious that the unit operating costs decrease with increasing power (see Fig. 2) for all types of fuel.

The highest unit operating costs were obtained when using pellets, and the lowest when using coal. The analysis of the specific current costs for the operation of boiler houses using fuel chips is not indicative. These costs also depend too much on the distance of fuel transport due to its low energy density.

The results obtained allow us to conclude that the replacement of coal with fuel pellets is economically impractical, primarily due to significant transport costs. However, in some cases, the difference in unit costs for heat supply in coal and pellet boilers of

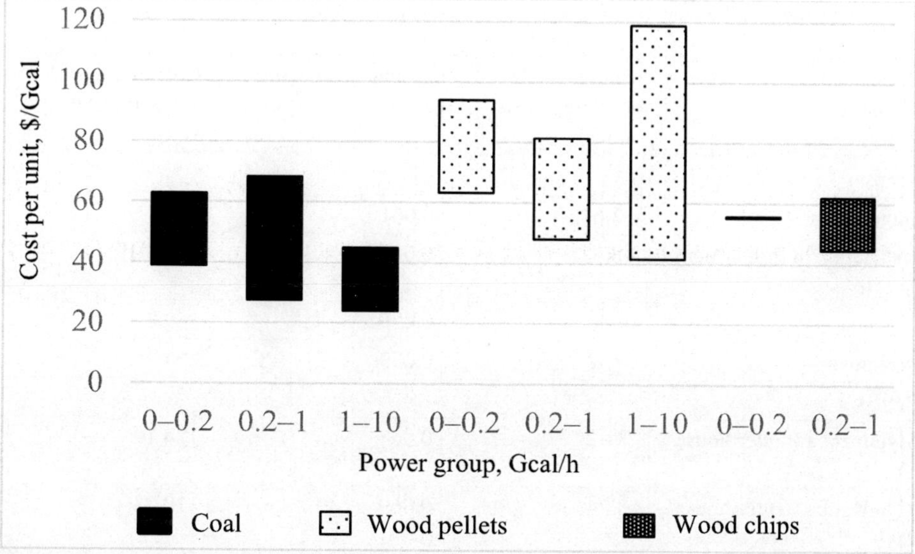

Fig. 2. Specific current costs for the production of thermal energy in boiler houses using coal and wood fuel.

the same capacity group is less than $ 15/Gcal. Accordingly, it is these boiler houses that are selected for ecological and economic analysis. These include 10 heat sources in the following localities: two boiler houses in the village of Kultuk, three boiler houses in Slyudyanka and one boiler house each in the village of Kudara, the urban-type settlements of Ust-Barguzin and Bolshaya Rechka, the village of Goryachinsk, the village of Yartsy and the village of Tvorogovo. Thus, for ecological and economic assessment, the use of fuel chips and pellets in 14 boiler houses operating in the coastal areas of Lake Baikal is considered.

The study showed that when coal is replaced with fuel chips in the village of Buguldeika and the urban-type settlement of Ust-Barguzin (boiler house 1 in Table 2), the unit operating costs decrease. This is due to a significant reduction in transport costs, despite the fact that the bulk density and calorific value of fuel chips is significantly lower than that of coal. The distances for transporting coal to these settlements are large, and fuel chips, on the contrary, are insignificant. Such substitution leads to a double effect – both ecological and economic.

The cost of reducing 1 ton of emissions when replacing coal with pellets depends on the qualitative characteristics of the coal used through the environmental component and on the transport factor through the economic one. For example, the highest cost of reduced emissions was in the village of Bolshaya Rechka – 477.15 $/t. This is due to the replacement of Borodinskii coal with eco-friendly quality characteristics (low ash content of 7.9% and sulfur content of 0.2%) with pellets requiring expensive transportation by road. The lowest in the calculations was the cost of reduced emissions in the village of Yartsy – 39.93 $/t. This is due to the rejection of Tugnuiskii coal with high ash content (23.3%) in favor of pellets, which are produced just 13 km from the village. On average,

Table 2. The cost of reduced emissions when replacing coal with wood fuel.

Settlement (Boiler house)	Change in unit operating costs, \$/Gcal	Change in specific emissions, t/Gcal	Cost of reduced emissions, \$/t
Chips			
Buguldeika	−7.61	0.061	−124.94
Ust-Barguzin (boiler house 1)	−8.52	0.082	−104.19
Gusikha	6.86	0.076	89.89
Adamovo	4.52	0.080	56.35
Pellets			
Slyudyanka (boiler house 1)	8.43	0.068	124.19
Slyudyanka (boiler house 2)	8.24	0.069	119.83
Slyudyanka (boiler house 3)	11.50	0.068	168.18
Kultuk (boiler house 1)	12.61	0.081	154.84
Kultuk (boiler house 2)	8.62	0.069	125.05
Ust-Barguzin (boiler house 2)	8.56	0.065	132.42
Goryachinsk	9.26	0.052	179.41
Yartsy	2.58	0.065	39.93
Bolshaya Rechka	12.00	0.025	477.15
Tvorogovo	13.77	0.083	165.32

the value of the cost of reducing emissions when replacing coal with pellets in the coastal areas of Lake Baikal varies on average from \$ 120–179/t and is typical for small and medium-capacity boiler houses, which currently use environmentally unfavorable coal of the Cheremkhovskii coal.

3.4 Additional Environmental Effects of Replacing Coal with Wood Fuel

The above-obtained ecological and economic estimates of the replacement of coal with chips and pellets in a number of boiler houses of the coastal territories of Lake Baikal can be supplemented with other environmental effects, since only emission of the pollutants (Emission) were considered when calculating the cost of reduced emissions, without taking into account the reduction of CO_2 emissions and the formation of SAW. Additional environmental effects from the replacement of coals with fuel chips and pellets are presented in Table 3.

Table 3. Environmental effects of replacing coal with wood fuel.

Fuel type, indicator	Pellets			Chips		
	Emission	CO_2	SAW	Emission	CO_2	SAW
Coal, t/year	1692.3	20305.7	1672.8	93.4	1252.1	89.0
Biofuels, t/year	92.6	1.2	138.9	5.3	0.2	7.2
Decrease, by t/year	1599.7	20304.5	1533.9	88.1	1251.9	81.8
Effect, ↓ on %	94.5	99.9	91.7	94.0	99.9	92.0

In general, the emission of the pollutants into the atmosphere is reduced by 17.6 times, and the formation of SAW is reduced by 12.4 times for 4 coal-fired boiler houses recommended for replacement with fuel chips. In total, when replacing coal with pellets, the reduction in emission of the pollutants into the atmosphere is estimated at 18.3 times, and the SAW–12 times.

Considering these 14 coal-fired boilers as a whole, potentially suitable and environmentally and economically justified for substitution with wood fuel, the effect for the coastal areas of Lake Baikal can be as follows: reduction of emission of the pollutants into the atmosphere by 1.7 thousand t/year, complete reduction of sulfur oxide emissions, reduction of the formation of SAW by 1.6 thou tons, reduction of CO_2 by 21.6 thou t/year.

4 Conclusion

The use of wood waste from timber industry for energy purposes has a double effect: ecological (in terms of waste disposal) and economic (the producer receives additional income from the sale or own consumption of the secondary product, and the consumer receives eco-friendly fuel at a relatively low price).

However, due to the features of wood fuel, when assessing ecological and economic indicators, it is necessary to take into account the transport factor and the capacity of the heat source. The use of the method of ecological and economic assessment made it possible to recommend 14 boiler houses potentially suitable for replacing coal with wood fuel from 54 boiler houses considered in the coastal areas of Lake Baikal. Four of these are recommended for burning of fuel chips, and 10 for the use of pellets.

The study showed that during the production and transportation of biofuels, CO_2 emissions are ten times higher than during open-pit coal mining and transportation. However, after taking into account the entire production–transportation–combustion cycle, CO_2 emissions for biofuels turned out to be significantly higher than for coal. In general, replacing coal with wood fuel in only 14 boiler houses at the coast of Lake Baikal will reduce the emission of pollutants by 94%, greenhouse gas emissions by 99.9%, and the formation of SAW by 92–91%.

Acknowledgements. The research was carried out under State Assignment Project (no. FWEU-2021-0004) of the Fundamental Research Program of Russian Federation 2021–2030 using the resources of the High-Temperature Circuit Multi-Access Research Center (Ministry of Science and Higher Education of the Russian Federation, project no 13.CKP.21.0038).

References

1. Sobolev, S., Trofimuk, V.: Krasnoyarsk Territory raw materials use complexity problems at wood-processing enterprises. Success Mod. Nat. Sci. **4**, 168–171 (2004). (in Russian), https://natural-sciences.ru/ru/article/view?id=12611. Accessed 13 Mar 2022
2. IEA: CO_2 emissions from fuel combustion (2000–2018). https://www.iea.org/data-and-statistics/charts/CO2-emissions-from-fuel-combustion-2000-2018. Accessed 10 Jun 2022
3. The Paris Agreement. United Nations (2015). https://unfccc.int/process-and-meetings/the-paris-agreement/the-paris-agreement. Accessed 07 Sep 2022
4. Friedmann, J., Fan, Z., Byrum, Z., Ochu, E., Bhardwaj, A., Sheerazi, H.: Levelized cost of carbon abatement: an improved cost-assessment methodology for a net-zero emissions world. Columbia University CGEP, New York (2020). https://www.energypolicy.columbia.edu/sites/default/files/file-uploads/LCCA_CGEP-Report_101620.pdf. Accessed 07 Sep 2022
5. Baker, E.D., Khatami, S.N.: The levelized cost of carbon: a practical, if imperfect, method to compare CO_2 abatement projects. Clim. Policy **19**(9), 1132–1143 (2019)
6. Vogt-Schilb, A., Meunier, G., Hallegatte, S.: When starting with the most expensive option makes sense: optimal timing, cost and sectoral allocation of abatement investment. Environ. Econ. Manage. **88**, 210–233 (2018)
7. Zia, H.: LCCA and environmental impact of buildings. In: Reference Module in Materials Science and Materials Engineering, pp. 133–143 (2020)
8. A technique for determining the gross emissions of pollutants into the atmosphere from TPP boiler plants. https://meganorm.ru/Data2/1/4294844/4294844842.htm. Accessed 07 Sep 2022. (in Russian)
9. A technique for determination of pollutant emissions from fuel combustion in the boilers with a capacity below 30 tons of steam per hour or 20 Gcal per hour. https://meganorm.ru/Data2/1/4294849/4294849657.htm. Accessed 07 Sep 2022. (in Russian)
10. Methodological letter No. 335/33–07 by the Atmosfera Research Institute "On calculating pollutant emissions into the atmosphere as per" A technique for determination of pollutant emissions from fuel combustion in the boilers with a capacity below 30 tons of steam per hour or 20 Gcal per hour. https://portal.tpu.ru/SHARED/a/ANL/work2/Tab1/Tab2/Методическое%20письмо%20НИИ%20Атмос.pdf. Accessed 10 Feb 2022. (in Russian)
11. Order No. 300: On the scope of guidelines and fees for the number of meetings of coverage of gas organizations united by economic and other activities in the Russian Federation. https://www.rosteplo.ru/Npb_files/npb_shablon.php?id=1740. Accessed 07 Sep 2022. (in Russian)
12. Good Practice guidance and uncertainty management in national greenhouse gas inventories. https://www.ipcc-nggip.iges.or.jp/public/gp/english/2_Energy.pdf. Accessed 07 Sep 2022. (in Russian)
13. The ENplus GHG calculator. https://www.enplus-pellets.eu/en-in/component/attachments/?task=download&id=119:GHG-Calculator_pellets_v1. Accessed 07 Sep 2022. (in Russian)
14. On the protection of Lake Baikal. Russian Federation, Federal Law No. 94-FZ (1999). http://www.kremlin.ru/acts/bank/13787. Accessed 07 Sep 2022. (in Russian)
15. Examination of nominations of cultural and natural properties to the World Heritage List and List of World Heritage in Danger. 20th extraordinary session of the Bureau. UNESCO World Heritage Centre (1996). https://whc.unesco.org/archive/1996/whc-96-conf201-9e.pdf. Accessed 07 Sep 2022 (in Russian)

16. Saneev, B.G., Ivanova, I., Maysyuk, E.P., Tuguzova, T.F., Ivanov, R.A.: The energy infrastructure of the central ecological zone: impact on the natural environment and ways to reduce it. Geography Nat. Resour. **5**, 218–224 (2016). (in Russian)
17. Buss, J., Mansuy, N., Laganière, J., Persson, D.: Greenhouse gas mitigation potential of replacing diesel fuel with wood-based bioenergy in an artic Indigenous community: a pilot study in Fort McPherson. Biomass Bioener. **159**(59), 106367 (2022)
18. Buss, J., Mansuy, N., Madrali, S.: De-risking wood-based bioenergy development in remote and indigenous communities in Canada. Energies **14**, 2603 (2021)

Demographic Potential of Agglomerations of Siberia: Influence of the Factor of Population Density

Natalia V. Emelianova, Yulia V. Antipina(✉), and Olga V. Valeeva

V.B. Sochava Institute of Geography, Siberian Branch, Russian Academy of Sciences, ul. Ulan-Batorskaya, 1, 664033 Irkutsk, Russia
Juliaav9@mail.ru

Abstract. We examine the main factors influencing the dynamics of population density in urban agglomerations of the Siberian Federal District, based on a case study of the Krasnoyarsk and Irkutsk agglomerations, among which there are the processes of natural and migration movement of population. It is shown that the contribution of migration to an increase in the population size prevails over the natural population increase in all regions of agglomerations. However, there occurs an intra-agglomeration differentiation with respect to the demographic potential. The Krasnoyarsk agglomeration shows a more stable natural population increase. On the other hand, Irkutsk exhibits a less favorable tendency; a stable increase in population size is characteristic for the nearest suburban zone only. This paper considers different effects from the factor of high and low population size on the population growth rates for different structural elements of urban agglomerations of Siberia: center cities, satellite cities, and the suburban zone. It is shown that more compact urbanized territories, such as center cities of agglomerations, the rates of population increase are limited to crowding, unlike the suburban zone where the population increase is directly proportional to the population size. The population increase of satellite cities is more dependent on the distance to large center cities of Siberian agglomerations and is less dependent on population density.

Keywords: Agglomerations of Siberia · Demographic potential · Dynamics of population size

1 Introduction

Demographic development of an urban agglomeration implies a change in population size and structure within particular spatial boundaries. In Russia, the demographic trends and the associated problems of population density relate to issues of national security [1]. In this respect, the population dynamics, local development and socio-economic differences are closely interrelated and call for a long-term investigation into representative socio-economic data at different geographical scales.

Siberia shows a considerable polarization of its territories according to the core-periphery principle as well as divergent trends n population size changes. Low indicators

I. N. Vladimirov et al. (Eds.): RERSDNA 2022, SPEES, pp. 36–45, 2023.
https://doi.org/10.1007/978-3-031-28978-1_5

of development, population density and socio-economic well-being of the peripherical regions of Siberia are becoming responsible for the unstable dynamics of population and general depopulation [2].

Concentration in cities and compactness of settlements are the cause of a deceleration in population growth as compared with urban conditions where the population density is below this threshold value and which show relatively higher growth rates [3–5]. Rural okrugs with a particularly low population density are experiencing negative rates of population increase more frequently than rural okrugs with a higher concentration of stable population [6]. An especially low population density and a constant population loss are indicators of a depopulation. A comparative analysis of the demographic dynamics for the last decade (2010–2020) appears to be suitable for determining threshold density values which are more likely to influence a further population dynamics of the Siberian agglomerations and to lead to positive (or negative) growth rates, both in urban settings and in rural areas.

A large number of factors were suggested, which influence the spatial distribution of stable population in Siberia in conditions of both high and low density [7–10]. While a compact urbanization, caused by internal migration (from rural to urban settlements), was associated with a high population density, mobility of inhabitants to suburban areas has recently led to an intermittent expansion of cities, which contributed to economic growth of suburban regions.

Suburbanization has influenced the structure of large cities and agglomerations by determining a deceleration of the population growth in many urban centers, and the unexpected acceleration in peripheral districts [11]. Based on these preconditions, a more careful investigation into the mechanisms of population growth that are density dependent in the urban and rural areas in the Siberian agglomerations can provide a basic integrated knowledge of the obvious or hidden processes of urban concentration/deconcentration in large cities, and rural depopulation in the more marginal districts. This knowledge also indicates differences in the demographic dynamics and economic polarization of regions with a high population dynamics and regions with depopulation.

2 Materials and Methods

Unlike foreign countries, no official statistics are kept concerning the agglomerations in Russia. P.M. Polyan and T.I. Selivanov developed a list of Russian agglomerations, including the Krasnoyarsk and Irkutsk agglomerations [12]. According to the Russian methodology, a group of settlements is defined as an urban agglomeration (UA) if the population of the largest core cities forming it exceeds 100 thousand and if within a 1.5-h transport accessibility from it there exist at least two urban-type cities or settlements gravitating toward the center [13].

The Irkutsk agglomeration has existed in the south of Irkutsk oblast for more than 50 years; it forms part of the Irkutsk-Cheremkhovo territorial production complex, which essentially determines its “prescriptive”, rather than natural, character of formation and development [14–16]. Several projects of the Irkutsk agglomeration were developed, such as the “Greater Irkutsk” [17] and “Baikal Ray” [18]. The concept of the “Greater Irkutsk” from the year 2007 implies uniting municipal districts around the city of Irkutsk, with a high population density remaining. Within the boundaries of the Irkutsk agglomeration, two suburban zones are singled out: the boundaries of the first zone include only the nearby satellite cities of Shelekhov and Angarsk, and the second zone includes remote satellite cities (Usolye-Sibirskoe, and the settlement of Listvyanka). The composition of the "Baikal Ray" project 2016 was reduced, with the new concept including only Irkutsk, Angarsk, Shelekhov, Baikalsk, and some of the settlements of the Irkutskii, Shelekhovskii and Slyudyanskii districts.

According to the "Krasnoyarsk 2020" concept, the boundaries of the Krasnoyarsk agglomeration coincide with the boundaries of the municipal districts adjacent to Krasnoyarsk [19]. In the project of the Krasnoyarsk urban agglomeration, the boundary of the nearby ring of the zone of influence of the center city is marked at the distance of 60 km from the center city, or within a 1.5-h transport accessibility.

3 Results and Discussion

The population size of the urban agglomerations of Siberia for the period from 2011 to 2020 shows a positive dynamics, with an intra-agglomeration differentiation associated with a continuous process of movement of the population, both mechanical and natural, which is determined by the rates of territorial development.

Most of the population growth in the Krasnoyarsk agglomeration corresponds to the Emelyanovskii district which is nearest to the capital city of the region and is characterized by a well developed network of rural settlements and the emerging network of cottage and dacha settlements. In 2020, the contribution of the Emelyanovskii district to the population growth of the Krasnoyarsk agglomeration made up 73%. The remaining percentage of the population growth corresponds to the Berezovskii district (14.6%), and to the cities of Sosnovoborsk (11.1%) and Divnogorsk (1.3%). A decline in population is observed in the Manskii and Sukhobuzimskii districts – they are agricultural areas located on the periphery of the agglomeration. The population size in the city of Krasnoyarsk was steadily increasing until 2019, followed by a decline thereafter.

In the Irkutsk urban agglomeration, the population growth is ensured by the Irkutskii district which from year to year is stepping up the pace of population growth. Thus, during the last year the population size increased by a factor of 1.1 (from 129.6 thou to 134.8 thou people). A positive dynamics is also observed in the Shelekhovskii district. In the capital city of the region, the population growth gave way to a decline after 2017. The other districts of the agglomeration are losing their population because of the negative migration balance and a natural population loss.

The main factor influencing the population dynamics is represented by the processes of natural and migration movement of population. The contribution of migration to an increase in the population size is considerably higher when compared with the natural growth in all districts of the agglomerations.

As pointed out above, the population size in the agglomerations increases, with a concomitant gradual natural loss of population. In the Krasnoyarsk agglomeration before 2013, the reproduction of the population showed a positive tendency, whereas by the year 2014, none of the districts could ensure a natural growth of population. During the period from 2014 to 2017, in the Sukhobuzimskii and Manskii districts the migration growth of population was overlapped by a natural outflow with the result that no increase in the population size occurred. Most of the migration growth of population of the agglomeration falls on the Emelyanovskii district (during the period 2011–2019, it increased from 11 to до 58.2%). During 2011–2018, in the city of Krasnoyarsk the favorable demographic situation with a positive natural and migration growth of the population was gradually decelerating its pace with the result that beginning in 2019 the center of the agglomeration fell into a demographic crisis: the population size decreased because of an abrupt migration decline (from 164 to −22 ‰) as well as due to the negative dynamics of the reproduction process: the mortality rates exceeded the birthrate. Such a phenomenon can be attributed to the preference of the inhabitants of the capital city to move to the suburban zone as well as because elderly people are dominant in the city, due to which the mortality rate has increased.

In the Irkutsk agglomeration, the population size is increasing in the suburban zone, primarily in the Irkutskii and Shelekhovskii districts which are the most favorable as regards demographic development: over the course of the last decade they have retained a positive natural and migration growth of population. Furthermore, most of the population is concentrated nearby the city of Irkutsk. Before 2018, in the center of the agglomeration the population growth was ensured by a natural movement, which could lead to a more clearly pronounced increase in the population size but was restrained by the migration outflow. For the period under consideration the city of Usolye-Sibirskoe has been losing its population, both due to the negative reproduction process (from 0.1 to −6.4 ‰) and because of a migration loss of the population (from −116 to −66 ‰).

Thus the intra-agglomeration differentiation with respect to demographic development is indicative of a rapid population growth in the nearby suburban zone which attracts the population from the center. Core cities are gradually decreasing the growth rates of the demographic potential. Districts and satellite cities at a considerable distance from the center are experiencing the effect of negative demographic processes.

The agglomerations of Siberia are highly differentiated territories; therefore, for a comparative study it is more convenient to examine the density indicator in interrelationship with the population dynamics separately for different structural elements (center city of an agglomeration, satellite cities, and suburban territories) which differ greatly in their demographic trends.

The unevenness of population density distribution directly affects the growth rates of inhabitants in the okrugs and districts of the cities. Urban okrugs and districts have a complex irregular population dynamics. Analysis is complicated by the inequality of cities with regards to the population growth and size and the average dynamics as well as by an additional factor affecting the attractiveness of separate districts of the city, including the cost of housing, the existence of vacant areas for development, the level of crime, availability of social service facilities, ecological status etc. Nevertheless, the research revealed a trend common to all cities. In each separate center city of the agglomerations, a low and medium population density of up to 5000 pers/km^2 is most attractive for inhabitants. Districts with a very high and very low density show a lesser population growth.

According to the summary diagram for all the objects under consideration, the Siberian center cities of the agglomeration correlate in general with the European trends

[3]: the urban districts with a very high population density show a lower population growth. That is, the factor of negative dependence of the population size and dynamics operates in such districts. The main peak of the positive population dynamics of the center cities in Siberia's agglomerations, according to the summary diagram, corresponds to the population density of okrugs (districts), 2000 pers/km^2 (see Fig. 1). Upon reaching 4000 pers/km^2, the population growth is not so high anymore. Krasnoyarsk shows the most clearly pronounced differentiation with respect to the population growth in the districts/okrugs. This correlates with the general high rates of population growth in the city. It can be suggested that the higher the population growth will be in the cities of Siberia, the larger will be the degree of differentiation of the urban districts with respect to the population growth. However, data on the agglomerations are inadequate to determine the correlation coefficient.

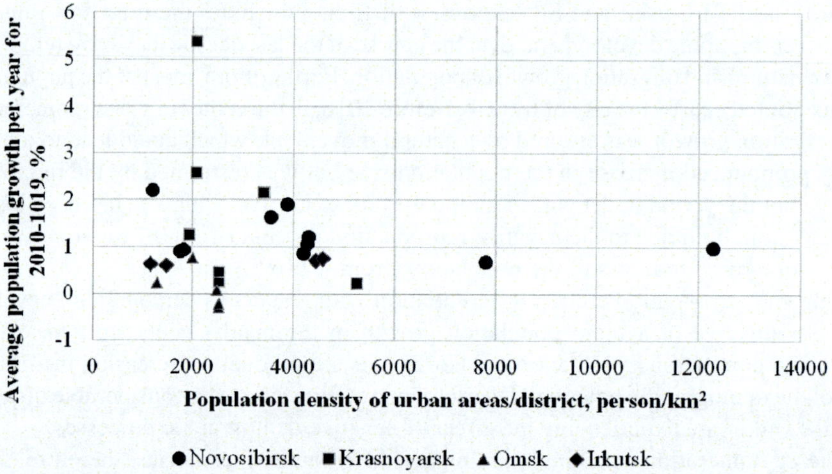

Fig. 1. Diagram of the dependence of population dynamics on density by urban districts of the center cities of large Siberian agglomerations for 2010–2018, %.

Suburban areas of the center cities of Siberia's agglomerations are represented mostly by rural settlements and show different trends as compared with the urban districts of the center cities. The diagram presents data on urban and rural settlements forming part of municipalities nearby centre cities of the districts for 2010–2020 (see Fig. 2). The tendencies are difficult to assess because of the inequality of the Siberian agglomerations as regards the total population size and dynamics. However, in spite of all the differences among the objects under investigation, it is possible to identify a trend common to the suburbanized territories, i. e. the growth of population of urban and rural settlements is, in all cases, proportional to population density. A positive dependence of the degree of impact of the effect of density on economic development and on the size of the agglomeration itself can also be identified. For instance, in the rural and urban settlements of the suburban zone of the Krasnoyarsk agglomeration, the dependence of the population growth on the population size is higher than in the Irkutsk agglomeration. Accordingly, the correlation coefficients for the urban and rural settlements in the

Krasnoyarsk agglomeration (0.53) are higher than in the Irkutsk agglomeration (0.28). A summary analysis for all the regions under consideration shows the same trend: the growth of inhabitants of urban and rural settlements is proportional to the population size of these settlements. These data are also correlated with European studies into the dependence of the dynamics on density [3]. The total correlation coefficient for the suburbanized territories of the Siberian agglomerations is 0.35.

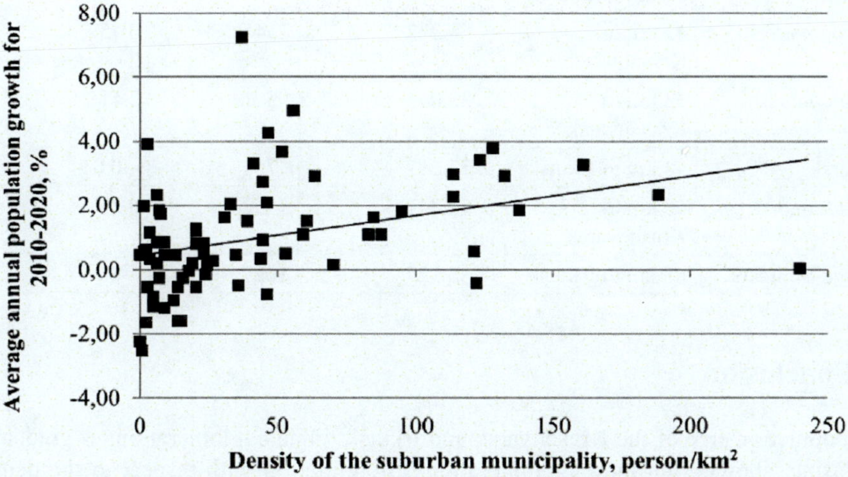

Fig. 2. Diagram of the dependence of population dynamics on density for urban and rural settlements in suburban areas of the large Siberian agglomerations: Novosibirsk, Krasnoyarsk, Omsk, and Irkutsk.

Data on satellite cities (and on large satellite workers' settlements with more than 10 thousand inhabitants) are sparse and therefore cannot be representative for in-depth analysis. However, some tendencies are relatively clearly pronounced. The population dynamics of the satellite cities (satellite workers' settlements) of the Siberian agglomerations (as well as their population size) is, to a greater extent, dependent on the distance to the center city (see Table 1) and, to a lesser extent, on population density. A correlation of the indicators of population growth and population size in satellite cities (satellite workers' settlements) with a coefficient of 0.13 is accounted for by the dependence of the two indicators on the proximity to the center cities. On average, more populous satellite cities (satellite workers' settlements) have more clearly pronounced positive dynamics then less populous cities.

In the Irkutsk agglomeration, the satellite cities show a relatively low population growth, and the remote cities are experiencing a population decline. The Krasnoyarsk agglomeration, with its center city of a moderate size where even sufficiently remote cities show a small growth, show the most stable population dynamics of the satellite cities.

Table 1. Dependence of the dynamics of the satellite cities' population on the distance to the cities-centers of the Krasnoyarsk and Irkutsk agglomerations.

Satellite cities of agglomerations	Distance from the satellite cities to the center of the agglomeration, km	Population as of 2010	Population density, person/km^2	Average population growth for 2010–2020, %
Shelekhov	23 km to Irkutsk	47.943	1542	0.10
Divnogorsk	41 km to Krasnoyarsk	28.272	1545	0.45
Sosnovoborsk	43 km to Krasnoyarsk	33.091	501	2.41
Angarsk	44 km to Irkutsk	233.567	764	−0.01
Zheleznogorsk	60 km to Krasnoyarsk	93.841	457	0.26
Usolye-Sibirskoe	88 km to Irkutsk	83.327	1027	−8.74

4 Conclusion

The population size of the Krasnoyarsk and Irkutsk urban agglomerations is gradually increasing, showing an intra-agglomeration differentiation with respect to the demographic potential. The core cities show a slowdown of the reproduction processes and migration growth of population. Nevertheless, the city of Krasnoyarsk is characterized by increased values of indicators of mechanical and natural movement, whereas the trends is less favorable for Irkutsk. A steady growth of the population size is characteristic for the nearby suburban zone which attracts the population from the center. The regions and satellite cities at a large distance from the center are experiencing the effect of negative demographic processes.

The main trends of the Siberian agglomerations correspond to the European trends of the dependence of population increase on population size. In the center cities of the agglomerations, the population density and size can become a deterrent against the population increase in urban districts. The larger is the population size of urban and rural settlements of the suburban zone of agglomeration, the higher is the population growth in such settlements (see Fig. 3).

A difference from the European trends implies the characteristics of the distribution of these indicators in satellite cities in which the population dynamics, and also their population size are, to a greater extent, dependent on the distance from the center city. Thus, in general, this suggests a tendency of the population toward the more populous and more densely populated territories of Siberia's agglomerations but without exceeding certain limits of density.

The Siberian agglomerations are also characterized by a strong differentiation of the territories with respect to the density-dependent indicators of population dynamics. Specifically, an uneven distribution of these indicators is typical for the largest and economically developed agglomerations.

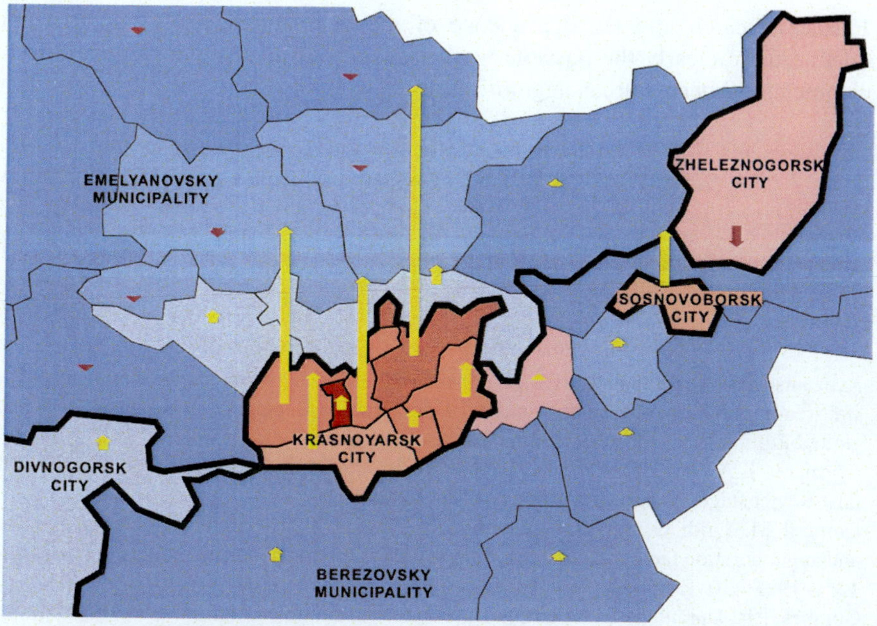

A. Krasnoyarsk urban agglomeration

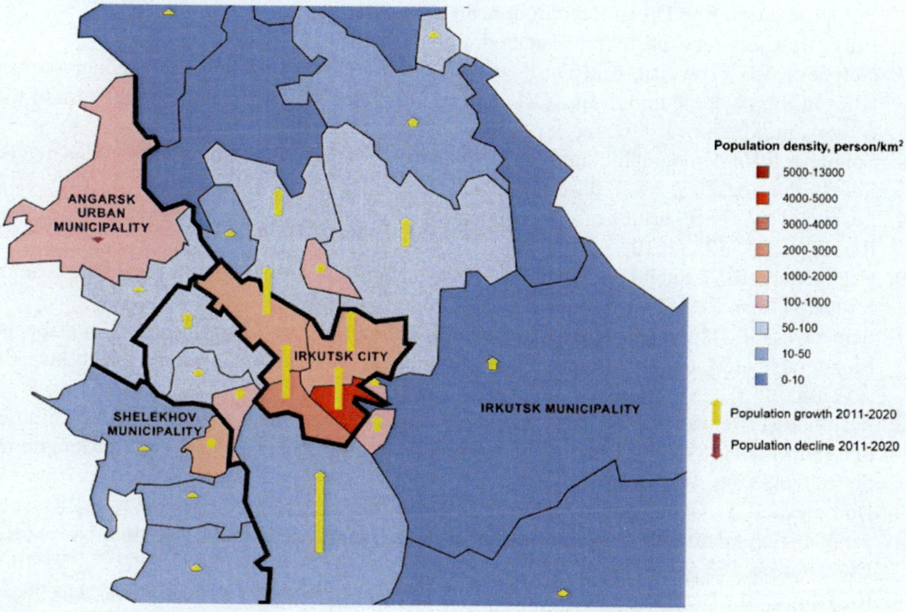

B. Irkutsk urban agglomeration

Fig. 3. Dependence of population dynamics on the population size of Siberian urban agglomerations: A. – Krasnoyarsk, B. – Irkutsk.

Further research into the dependence of the population density and dynamics will define more clearly the possible ways to work out new solutions regarding the development of Siberia's urban agglomerations.

Acknowledgments. This research was financially supported by a state assignment (state registration no. AAAA-A21-121012190019-9) and by a grant from RFBR and MECSS (project no. 20-55-44023).

References

1. Ryazantsev, S.V., Arkhangelskii, V.N., Vorobieva, O.D.: Demographic development of Russia: trends, forecasts, measures. National Demographic Report. In: Ryazantsev, S.V. (ed.) United Edition LLC, Moscow, 156 p. (2020)
2. Bezrukov, L.A.: Transformation of economic structures and population of Siberia at the post-Soviet stage. Geogr. Orir. Resurs. **4**, 25–36 (2020)
3. Ciommi, M., Egidi, G., Salvia, R., Cividino, S., Rontos, K., Salvati, L.: Population dynamics and agglomeration factors: a non-linear threshold estimation of density effects. Sustainability **12**, 1–19 (2020)
4. Combes, P.P., Duranton, G., Gobillon, L.: The identification of agglomeration economies. J. Econ. Geogr. **11**, 253–266 (2011)
5. Haase, A., Bernt, M., Großmann, K., Mykhnenko, V., Rink, D.: Varieties of shrinkage in European cities. Eur. Urban Reg. Stud. **2**, 86–102 (2016)
6. Price, D.: Carrying capacity reconsidered. Popul. Environ. **21**, 5–26 (1999)
7. Nefedova, T.G., Treyvish, A.I.: The theory of "differential urbanization" and the hierarchy of cities in Russia at the turn of the XXI century 1. Problems of urbanization at the turn of the century. In: Makhrova, A.G. (ed.), p. 7186. Oikumena, Smolensk (2002)
8. Soboleva, S.V.: Demographic situation in Siberia against the background of all-Russian trends. Soc. Probl. Reg. Dev. **2**(82), 98–115 (2014)
9. Lesyuta, N.V.: Peculiarities of development of urbanized territories of Siberia. Geog. Prir. Resurs. **1**, 119–123 (2010)
10. Grigorichev, K.V.: Suburban diversity: suburbanization in the Siberian Region (the case of Irkutsk). Urban Res. Pract. **2**, 7–23 (2016)
11. Semenov, V.T., Shtompel, N.E.: Sustainable Development of Megacities. Urban Aspects. Monograph. O.M. Beketov National University of Urban Economy in Kharkiv, Kharkiv, 339 p. (2014)
12. Polyan, P.M., Selivanova, T.I.: Urban agglomerations of Russia and new trends in the evolution of its network (1989–2002). In: Proceedings of the Russian Academy of Sciences, Geography Series, vol. 5, pp. 18–27 (2007)
13. Dorofeeva, L.A.: Geographical features of the formation of the suburban area of the Krasnoyarsk agglomeration. PhD thesis. V.P. Astafyev Krasnoyarsk State Pedagogical University, Krasnoyarsk, 155 p. (2018)
14. Korytny, L.M., Bardash, A.V., Bogdanov, V.N., Vorobyev, N.V.: Geographical aspects of substantiating the development of the greater Irkutsk agglomeration. Geogr. Prir. Resurs. **3**, 103–110 (2008)
15. Vorobyov, N.V., Emelyanova, N.V.: Irkutsk Urban Agglomeration: Features of Formation and Substantiation of Borders. V.B. Sochava Institute of Geography SB RAS Publishers, Irkutsk, 25 p. (2014)

16. Emelyanova, N.V., Geographic assessment of the interaction of cities within the Irkutsk agglomeration. PhD thesis. V.B. Sochava Institute of Geography SB RAS, Irkutsk, 187 p. (2013)
17. The concept of development of the Irkutsk agglomeration. FRRIO (2007)
18. Baikal ray: Project Baikal, vol. 43, pp. 34–36 (2014)
19. Government of the Krasnoyarsk Territory: Order of May 12, 2011, no. 326-r (On approval of the concept for the development of vocational education in the Krasnoyarsk Territory until 2020). https://docs.cntd.ru/document/985022197. Accessed 21 Oct 2021

Landscape-Ecological Zoning of a Territory as the Basis for Identifying Lands of the Ecological Framework of an Administrative District

Elena M. Klimina(✉) and Andrey V. Ostroukhov

Institute of Water and Ecological Problems, Far Eastern Branch, Russian Academy of Sciences, 56, ul. Dikopoltseva, 680000 Khabarovsk, Russia
{kliminaem,Ostran2004}@bk.ru

Abstract. A landscape and ecological zoning is carried out for the preservation of geosystems of high ecological value and for substantiating of their inclusion in ecological frameworks of territories of a different rank. This type of zoning is regarded as a combination of territorialy interconnected groups of geosystems in terms of the set of physical-geographical factors, and structural and functional features. For the Vaninskii and Sovetsko-Gavanskii municipal districts of Khabarovsk krai we identified geosystems of high ecological value at the regional level. The specially protected areas now make up a mere 26.8% of their total area, and this is inadequate to ensure their conservation. It is therefore worthwhile to include in the recommended ecological framework of municipal districts this category of geosystems with the necessary regulations of the modalities of nature management. As a special element of the ecological framework of the territory, the area of the conventionally unchanged and slightly modified geosystems with ecological functions of regional significance will additionally make up 26.3% (taking into consideration their inclusion in a number of other elements of the framework). As a result, the total proportion of lands in the ecological framework of the districts will be increased to 58.6% of their area, which corresponds to the optimal indicators of rational nature management in the subzone of the middle and southern taiga.

Keywords: landscape and ecological zoning · ecological framework of the territory · environment-forming functions · Vaninskii and Sovetsko-Gavanskii municipal districts · ecologically significant landscapes

1 Introduction

An important issue of development for intact territories is, as before, the problem of assessing the surviving potential of natural resources', including the ecological potential of a territory which ensures a stable functioning of geosystems. Within the framework of a municipal district, the solution of the problem of assessing ecological functions of

I. N. Vladimirov et al. (Eds.): RERSDNA 2022, SPEES, pp. 46–56, 2023.
https://doi.org/10.1007/978-3-031-28978-1_6

geosystems acquires a particular spatial manifestation. Furthermore, the nature conservation system of each district forms part of the regional system of the preservation of the natural environment and maintenance of the natural equilibrium which is most often formed as a different-level, multi-scale ecological framework of the territory (EFT).

The ecological (natural, landscape and ecological) framework of the territory is treated by most researchers as a system of territorial units possessing the intrinsic value and performing particular ecological functions [1–7]. In foreign science, it was reflected in the concept of "Ecological Networks" which is being actively developed around the world [8–12]. The application of the landscape approach in terms of territorial units o management (municipal districts) furnishes a means of using assessments of geosystems of a local level in combination with economic spatial units of planning and management (agricultural lands, forestries, etc.).

The economic conditions of the regions of new development determine the priority of the resource functions of geosystems, above all, concerning wood raw material, which are in many respects responsible for the ecological situation. For a forest region, such as Khabarovsk krai, the relevance of this study is associated with the lease (for logging activity) of most of the remaining large forest ranges, which can lead to the utilization of resources without assessing the significance of the environment-forming role of geosystems.

For substantiating the inclusion of ecologically valuable geosystems in EFT of different levels, we developed the procedure of landscape and ecological zoning (LEZ) based on combining territorially interconnected groups of geosystems according to the set of physical-geographical factors, and structural and functional features. A key element of the landscape approach that is used in this study involves taking into account the multifunctionality of forest landscapes, a close relationship of biotic and abiotic components and the influence of landscape conditions on methods of forest exploitation [1].

Within the boundaries of a municipal district, LEZ encompasses geosystems of the intra-landscape level, having different ecological functions of a local rank and higher, which should be taken into account in programs of both municipal and regional development. Thus the goal of this study is to identify geosystems of high ecological significance/value in the process of LEZ for the inclusion in EFT of municipal districts.

2 Objects and Methods

The objects of this investigation are represented by the Vaninskii and Sovetsko-Gavanskii municipal districts of Khabarovsk krai located on the eastern macroslope of Sikhote-Alin, with their area totaling 41.5 thou km^2, or 37.3% of the area of the mountain system. These districts are characterized by the dominant position of the transport, forestries and fisheries sectors in the economy. A significant part of nature management on the territory of the districts is associated with forest exploitation. Thus, as of 2021, merchantable forests account for 71.5% of the territory of the districts (including 48.6% on lease). Of 2282.7 thou ha of forests leased for logging, only 1108.9 thou ha (48.6%) obtained certification within the PEFC Program (Program for the Endorsement of Forest Certification schemes), and none of the loggers received a certificate within the program of FSC-certified timber compartments [13]. It is significant that of the 66 compartments that were referred previously to intact forest territories, with their area measuring 365.7 thou ha, there remained only 47 (268.6 thou ha, or 26.6%).

Source material: the landscape map of Northern Sikhote-Alin at a scale of 1:200 000, Earth remote sensing data of medium spatial resolution (Landsat 3, 5, 7, 8) for the time interval from to 2020, field material, information on forests on the website of the Ministry of Forestry and Timber Processing of Khabarovsk krai [14], data on forests of high conservation value of Khabarovsk krai [14], etc. A processing of material was carried out in the ArcGIS 10.5 software. Use as also made of statistical methods, the method of comparative analysis and landscape and ecological zoning. LEZ included several stages. Initially, the inventory landscape map was used to reveal the priority environment-forming functions of geosystems which subsequently determined identification of ecologically significant geosystems of regional and local level. Next, criteria were developed to serve as the basis for identification of landscape and ecological zones.

3 Results and Discussion

Analysis of the functional role of geosystems drew on the following statements. In the first place, almost the entire territory of the districts refers to the state forestry fund lands and is the zone of concentration of timber and non-timber resources, and hunting grounds (a high resource-forming role of geosystems). Second, the environment-forming significance of many geosystems in the districts at the level not below the regional level is determined by their location on Sikhote-Alin as the territory of maximum biodiversity in the region. And, third, the environment-forming functions are treated as the priority functions for the preservation of geosystems and their separate components. The level of significance of the functions reflects the role in the system of ranking: regional – for the preservation of geosystems of a high hierarchical rank (landscape areas and regions), and local – for particular geosystems (types of stows).

In the course of the landscape and ecological zoning, six categories of geosystems were identified (according to the degree of modification, and the levels of significance of the environment-forming functions, each of which combines several groups of landscapes [15]. In the Vaninskii and Sovetsko-Gavanskii districts, the largest area (52.8%) is occupied by conventionally unchanged geosystems with ecological functions of regional significance. For moderately modified geosystems (about 20.5% of the area of the districts), the priority belongs both to the environment-forming functions and to the functions of environment recovery. The most modified geosystems occupy 28.3% of the area.

Intensive economic development of this territory is determined by the proximity of timber resource bases to the major ports of Khabarovsk krai, and by the existence of the motor road and the railroad, which made it possible to organize a large Vaninski–Sovetsko-Gavanskii industrial hub. Of course, this contributed to decreasing the ecologo-functional significance of geosystems. Analysis of the effects of fires and felling for the period from 1975 to 2010 revealed large areas of burned-over areas making up 19.4% of the area of the two districts. By the year 2020, recovery of forest vegetation occupied 54.5% of the territory affected by fires. Over the last decade, according to results from analyzing remotely sensing data for 2011–2020, the annual increase in the burned-over areas made up 1.15%, or 4.8 thou ha, which is by a factor of 5 less than this indicator for the period 1975–2010 (23.1 thou ha).

Changes in the cutover areas, on the contrary, showed their positive dynamics: the mean annual increase amounted to 5.3 thou ha for the period 1975–2010 and to 7.3 thou ha for the period 2011–2020. The spatial pattern of distribution of cutover areas reflects a shift by the year 2020 toward hard-to-reach middle mountains and to the coast of the Strait of Tartary. While during 1975–2010 the middle mountains accounted for 2.1% of the cutover area, during 2011–2020 it reached 5.9%. A similar tendency of development of the hard-to-reach territories is characteristic for the whole of Northern Sikhote-Alin in general. Large areas of fire-damaged forest (the fires of 1998) are involved in the process of intense forest recovery, as a result of which there arise young forests. However, significant areas of volcanic plateaus that were repeatedly affected by fires, show the formation of long-lasting derivative tall grass-shrub communities.

One of the problems of conservation of highly significant geosystems is their low and uneven representation in Specially Protected Natural Areas (SPNA). Only 5% of the area of SPNA correspond to the mid-mountain landscapes with subgoletz communities, and 38.5% refer to geosystems of the volcanic plateaus, elevated plains and intermountain depressions with southern-taiga dark-coniferous and light-coniferous communities. The largest proportion, 62%, corresponds to geosystems of mountain rivers with mountain-valley coniferous s and broad-leaved forests, which is associated with the existence of ichthyological zakazniks and the Khutinskii ecological corridor (see Fig. 1). On the whole, the level of representation in SPNA of all highly valuable geosystems of regional significance makes up 26.8%, which is inadequate for them to perform environment-forming functions of such a level. Therefore, the formation of EFT of municipal districts is, as before, currently relevant and serves as the key component of landscape planning. In our opinion, the geosystems that remain beyond the structure of the modern SPNA corresponding to the key landscape territories of a regional rank would be appropriate for inclusion in EFT of municipal districts in order to establish the necessary modalities of utilization (see Figs. 1 and 2). Such a step reflects the flexibility of the EFT structure, because currently the most demanded resources are concentrated within these territories.

For instance, 75.4% of the landscape of low mountains and volcanic plateaus with mountain-taiga fir-spruce, larch and spruce-larch forests are on lease for purposes of forest exploitation. The inclusion in the composition of the ecological framework implies the need to use the more benign technologies and introduce proper control over the compliance with legislation. One of the forms of imposition of restrictions on nature management would be the inclusion of these geosystems in Special Protection Forest Areas (SPFA) when developing plans for industrial development of the territory.

Table 1 presents the main elements of EFT of the Vaninskii and Sovetsko-Gavanskii districts. It is based on the existing SPNA of different types and protected zones of natural and anthropogenic objects (see Fig. 1) with a total area of 858.1 thou ha (20.5% of EFT). The category of unchanged and slightly modified geosystems with the functions of a regional level as identified in the course of the landscape and ecological zoning is defined in EFT as elements of the regulated modality (Table 1, EFT element 2.8). Seven groups of geosystems that were identified in greater detail are presented in Fig. 2 and in Table 2, showing their brief description, the main functions of the ecologically significant

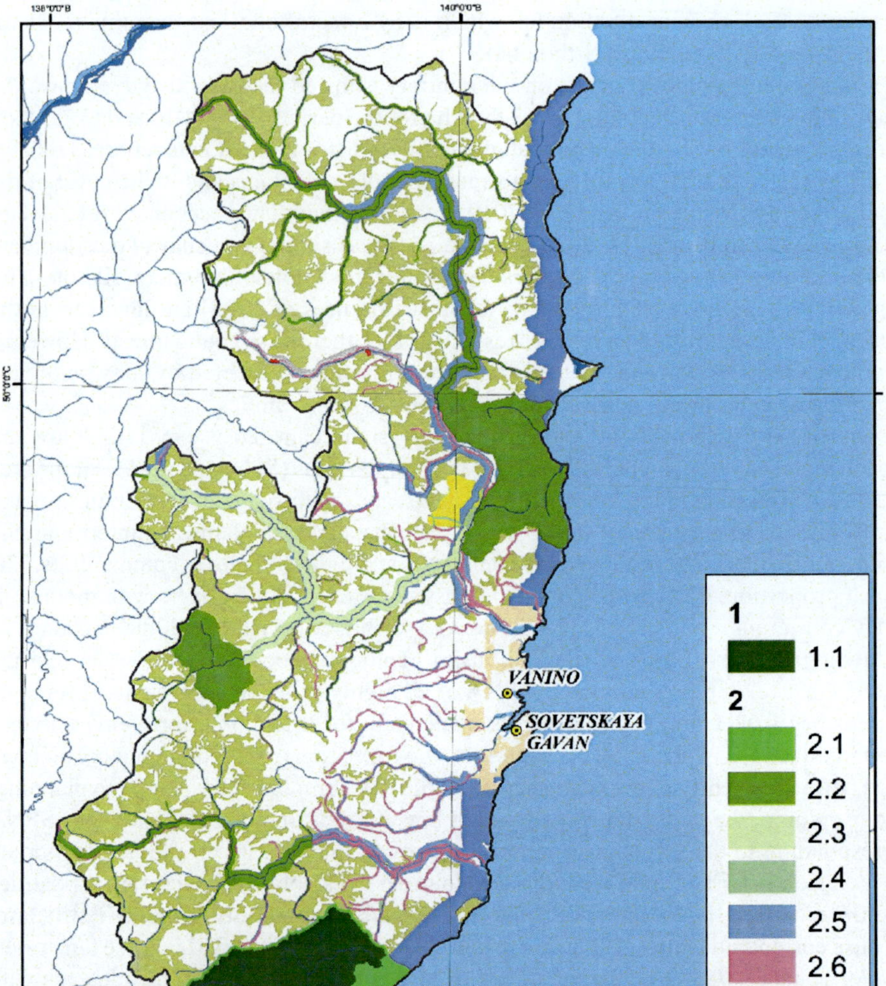

Fig. 1. Elements of the ecological framework of the Vaninskii and Sovetsko-Gavanskii districts (for symbols, see Table 1).

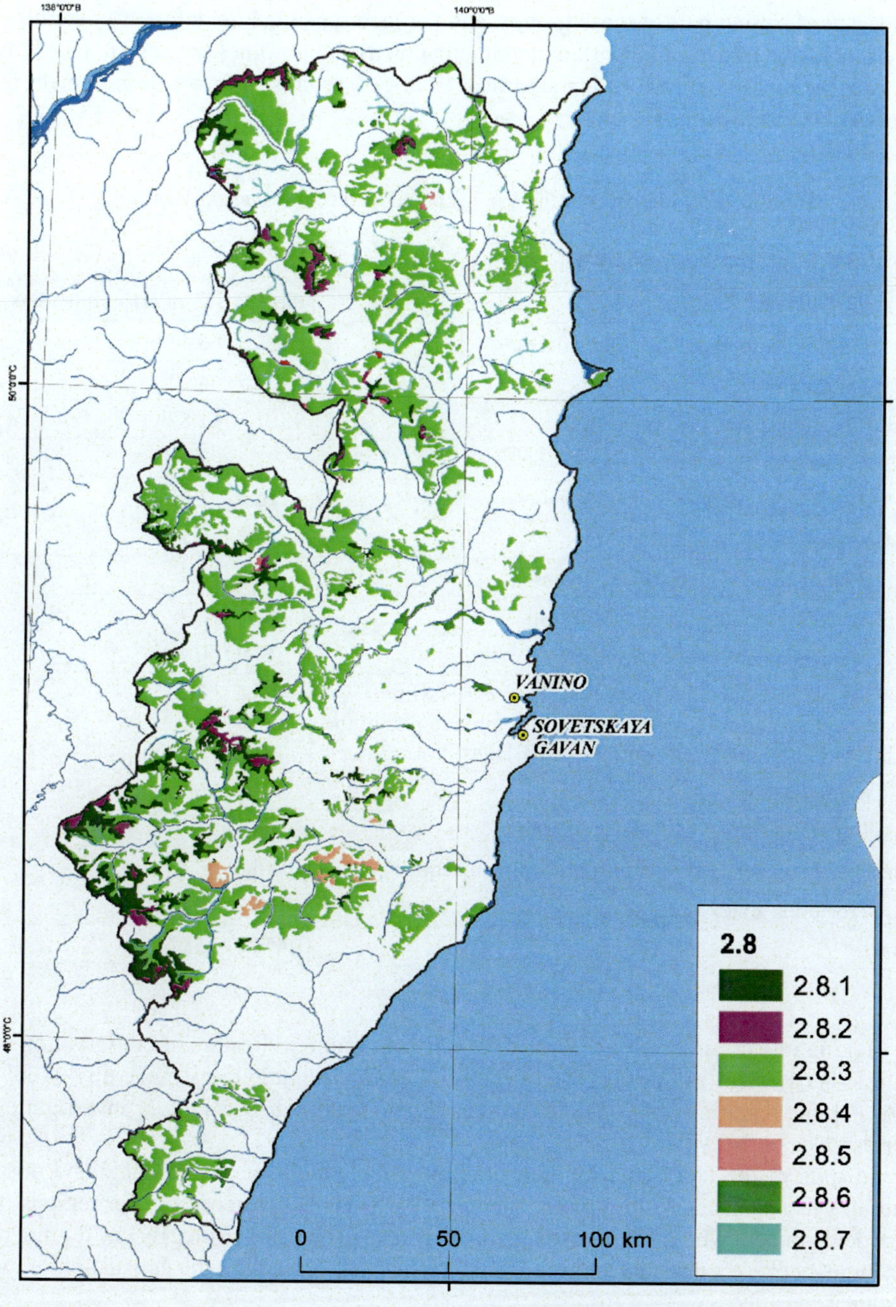

Fig. 2. Unchanged and slightly modified landscapes with ecological functions of the regional level (not included in the other more significant categories of EFT lands) (for symbols, see Table 2).

landscapes forming part of these groups, and the recommended modality of utilization. Their inclusion in the framework of a municipal district determines the need to take into account the environment-forming significance of geosystems in future development in a particular forest compartment.

Table 1. EFT elements of the Vaninskii and Sovetsko-Gavanskii districts.

Modalities of nature management and elements of the ecological framework	Area	
	thou ha	% of area of districts
1. Specially protected: 1.1. Zapovedniks	267.3	6.4
2. Regulated:	1927.5	46.3
2.1. Protection zones of zapovedniks	78.1	1.87
2.2. Zakazniks	3797	9.1
2.3. Ecological corridors	126.6	3.0
2.4. Natural monuments	6.4	0.1
2.5. Prohibited belts along water bodies	93.8	2.2
2.6. Spawning-protection belts of forests	141.0	3.4
2.7. Medical rehabilitation localities	14.7	0.4
2.8. Unchanged and slightly modified landscapes with ecological functions of regional level (not forming part of the other, more significant categories of EFT lands)	1100.4	26.3
3. Compensation	255.2	6.1
3.1. Buffer zones	39.9	1.0
3.2. Protection zones along transport communications	16.6	0.4
3.3. Protection zones along sea coasts	185.7	4.4
Total:	2450.3	58.6

Taking into consideration that the necessary minimum indicators of conservation of natural geosystems in the subzone of the southern and middle taiga make up 40–60% [16, 17], the resulting area of EFT equal to 58.6% for the districts under investigation corresponds to this value.

In spite of the fact that a number of authors include the post-pyrogenic and post-cutting geosystems as restoration elements of EFT, this is not necessary for regions of new development where nature management is focused on using primarily one dominant resource, because a targeted anthropogenic impact will cease with the exhaustion of the resource.

Table 2. Unchanged and slightly modified landscapes with ecological functions of a regional level.

Landscapes	Main functions	Modality of nature management	Area	
			thou ha	% of area of districts
2.8.1. Mid-mountain moderately and heavily dissected mountain - taiga, with fir-spruce, larch and spruce-larch forests	Landscape-forming, erosion-stabilizing and water-regulating, soil-protective	Protection, protective, strictly regulated	143.2	3.4
2.8.2. Mid-mountain with subgoletz communities	Landscape-forming, erosion-stabilizing and water-regulating	Protection, protective	39.1	0.94
2.8.3 Of low mountains and volcanic plateaus of a different degree of dissection, with mountain-taiga fir-spruce, larch and spruce-larch forests	Erosion-stabilizing and water-regulating, soil-protective	Preservation of the existing modalities with an enhancement of the protection modality	839.7	20.3
2.8.4. Of low mountains and volcanic plateaus of a different degree of dissection, with Siberian stone pine-spruce and Siber3ian stone pine-broad-leaved forests	Landscape-forming, erosion-stabilizing, runoff-forming, water-regulating	Protection, protective, strictly regulated	14.8	0.4
2.8.5. Low-mountain, with subgoletz communities	Erosion-stabilizing, water-regulating (in the north – permafrost-stabilizing)	Protection, protective	2.7	0.1
2.8.6. Of elevated plains and intramountain depressions, and of volcanic plateaus with southern-taiga dark-coniferous forests	Landscape-forming, biota-stabilizing, water-regulating, runoff-forming	Regulated, in combination with protective and protection	13.9	0.33

(*continued*)

Table 2. (*continued*)

Landscapes	Main functions	Modality of nature management	Area	
			thou ha	% of area of districts
2.8.7. Of floodplains of mountain rivers, with mountain-valley spruce-fir and larch, deciduous (with inclusion of coniferous and broad-leaved species) forests	Landscape-forming, water-regulating and water-protective, biot-stabilizing	Protection, protective	33.8	0.8
Total:			1087.4	26.3

4 Conclusion

Thus the landscape and ecological zoning revealed a negative annual dynamics of the area of undisturbed geosystems, which is due to a sable increase in logging operations – they are encroaching on hard-to-reach areas with highly vulnerable, ecologically valuable natural complexes.

For the ecological frameworks of the municipal districts we identified geosystems of high ecological value of a regional rank (47.5% of the area of the districts). And only 26.8% of these geosystems form part of the existing SPNA, which is inadequate for maintenance of an optimal functioning. In the structure of SPNA, they are represented unevenly: 24.2% of them correspond to the mountain-taiga low mountains (Botchinskii zapovednik, and Tumninskii and Mopau zakazniks), about 66% – to geosystems of the mountain floodplain (ecological corridors, and ichthyological zakazniks), and up to 10% – to the mountain-taiga middle mountains.

As part of the research it was found that 52.8% of the area of the districts correspond to the conventionally unchanged and slightly modified landscapes of regional and local significance. In EFT, they refer to lands with a regulated modality where their proportion makes up 26.3% with due regard for their inclusion in a number of other elements of the framework. On the basis of the results obtained, it is necessary to develop fractional elements of the framework for a deeper and comprehensive account of the regional and local value of particular geosystems. This will serve as an additional factor for sustainability of the natural environment and the preservation of its landscape diversity and integrity.

References

1. Khoroshev, A.V.: Mirovoi opyt primeneniya landshaftno-geograficheskogo podkhoda k planirovaniyu ustoichivogo lesopol'zovaniya (World experience in applying the landscape-geographical approach to planning sustainable forest management). Ustoichivoe lesopol'zovanie **1**(65), 13–20 (2021). (in Russian)

2. Mirzekhanova, Z.G.: Ekologicheskii karkas territorii v strategii ustoichivogo razvitiya: analiz podkhodov, naznachenie, soderzhanie (Ecological framework of the territory in the strategy of sustainable development: analysis of approaches, purpose, content). Geografiya i prirodnye resursy **2**, 154–158 (2001). (in Russian)
3. Danekina, V.N.: Landshaftno-ekologicheskoe planirovanie pribrezhnoi zony Yugo-Zapadnogo Kryma (Landscape and ecological planning of the coastal zone of the South-Western Crimea). InterKarto. InterGIS **1**(23), 277–285 (2017). (in Russian)
4. Batuev, A.R., Lopatkin, D.A.: Obosnovanie i kartografirovanie territorial'noi struktury ekologicheskogo karkasa regiona (Substantiation and mapping of the territorial structure of the ecological framework of the region). Izvestiya Irkutskogo gos. un-ta. Seriya «Nauki o Zemle» **1**(1), 56–75 (2008). (in Russian)
5. Mirzekhanova, Z.G., Ostroukhov, A.V.: Ekologicheskii karkas territorii v strategii regional'nogo razvitiya (na primere Vaninskogo raiona Khabarovskogo kraya) (Ecological framework of the territory in the regional development strategy (on the example of the Vaninsky district of Khabarovsk krai). Izvestiya RAN. Ser. Geograficheskaya **5**, 73–82 (2006). (in Russian)
6. Chibilev, M.A.A., Chibilev, A.A.: Sovremennoe sostoyanie i problemy modernizatsii prirodno-ekologicheskogo karkasa regionov stepnoi zony Evropeiskoi Rossii (Current state and problems of modernization of the natural and ecological framework of the regions of the steppe zone of European Russia). Yug Rossii: ekologiya, razvitie **1**(14), 117–125 (2019). (in Russian)
7. Myrzagalieva, Z.Z., Stanis, E.V.: Metodicheskie podkhody pri sozdanii ekologicheskogo karkasa Zapadno-Kazakhstanskoi oblasti (Methodical approaches to the creation of the ecological framework of the West Kazakhstan region). Vestnik Rossiiskogo un-ta druzhby narodov. Seriya: Ekologiya i bezopasnost' zhiznedeyatel'nosti **4**, 114–123 (2015). (in Russian)
8. Vasilijević, M., Pokrajac, S., Erg, B. (eds.): State of nature conservation systems in South-Eastern Europe, pp. 58. IUCN, Gland, Switzerland and Belgrade, Serbia (2018)
9. Vasilijević, M., Sanja, P., Boris, E.: Summary of national assessments of the state of nature conservation systems in South-Eastern Europe, pp. 84. IUCN, Gland, Switzerland and Belgrade, Serbia (2019)
10. Bennett, G., Kalemani, J.M.: Review of experience with ecological networks, corridors and buffer zones. Secretariat of the Convention on Biological Diversity, Montreal, Technical Series No. 23, 100 (2006)
11. Gutiérrez, N.S., Castillo, I.N., Niño, C.J.: Economic and ecological zoning proposal for La Roqueta Island. Mexico. ECORFAN Journal-Ecuador **3**, 28–39 (2016)
12. Hilty, J., et al.: Guidelines for conserving connectivity through ecological networks and corridors. Best Practice Protected. Area Guidelines Series 30. IUCN, Gland, Switzerland, 124 (2020)
13. Lesa vysokoi prirodookhrannoi tsennosti (LVPC) Khabarovskogo kraya (Forests of high conservation value (LVPC) of Khabarovsk krai). https://hcvf.ru/ru/maps/hcvf-khab, last accessed 2022/09/03. (in Russian)
14. Informatsiya o lesnom fonde Khabarovskogo kraya (Information about the forest fund of Khabarovsk krai). https://les.khabkrai.ru/Forest_Information/Obschedostupnaya-informaciya last accessed 2022/10/04. (in Russian)

15. Klimina, E.M.: Landshaftno-ekologicheskoe zonirovanie Severnogo Sikhote-Alinya (Landscape and ecological zoning of North Sikhote-Alin) in: Materialy XVI Soveshchaniya geografov Sibirii Dal'nego Vostoka. Vladivostok, September 2021, 142–145 (2021). (in Russian)
16. Shlotgauer, S.D.: Prirodookhrannye kompleksy i territorii (Nature conservation complexes and territories) in: Okhrana okruzhayushchei sredy I ratsional'noe prirodopol'zovanie Amursko-Komsomol'skogo TPK. DVO AN SSSR, Vladivostok, 136–143 (1988). (in Russian)
17. Reimers, N.F., Shtil'mark, F.R.: Osobookhranyaemye prirodnye territorii (Specially protected natural areas). Mysl', Moscow, 296 (1978). (in Russian)

Ecological Approach to Assessing the Causes and Consequences of the Winter Landslide on the Bureya Reservoir

Lubov M. Kondratyeva[1](✉), Zoya N. Litvinenko[1], Diana V. Andreeva[1], and Evgenia M. Golubeva[2]

[1] Institute of Water and Ecological Problems, Far Eastern Branch, Russian Academy of Sciences, 56, ul. Dikopoltseva, 680000 Khabarovsk, Russia
kondratyevalm@gmail.com

[2] Kosygin Institute of Tectonics and Geophysics, Far Eastern Branch, Russian Academy of Sciences, 65, ul. Kim Yu Chena, 680000 Khabarovsk, Russia

Abstract. The thawing of permafrost in the north of Siberia and the Far East is accompanied by hazardous geoogical phenomena (landslides, sink-holes, floods, etc.) which present a threat to the infrastructure of the region and lead to a change in the natural water quality. In December 2018, at the temperature of –32 °C, a giant landslide occurred on the shore of the Bureya reservoir, which cut off access of water to the Bureya hydroelectric power station. To restore the hydrologial regime, explosives were used, which also affected consideraby the composition of water in its interaction with crushed rocks. Risk factors were represneted by toxic volatile compounds (benzene, methanol, toluene, and xylene). As a result of destruction of soil coverof the coastal zone and biochemical processes at the interaction of water with rocks, considerable changes were observed in the behavior of a number of elements (As, Pb, Fe and Co). Ecological risks of change in the water composition are associted with input of a large volume of crushed rocks, vegetable debris, detonation products of explosives, and with an intensification of biogeochemical processes. They are accompanied by input of toxic organic matter, gases and elements into the water. It is necessary to conduct a comprehensive survey of the steep shores of the Bureya reservoir and organize a monitoring of the potential emission of methane in order to prevent hazardous natural phenomena associated with the thawing of permafrost.

Keywords: Landslide · Volatile organic compounds · Heavy metals

1 Introduction

Climate change is most often associated with emissions of greenhouse gases. However, an important ecological problem comes from the after-effects of the thawing of permafrost. In the northern regions of Siberia and the Far East, the thawing of permafrost in reecent years has been accompanied by hazardous geological phenomena (landslides, sink-holes, floods, etc.) which present a threat to the infrastructure of the reion and

I. N. Vladimirov et al. (Eds.): RERSDNA 2022, SPEES, pp. 57–67, 2023.
https://doi.org/10.1007/978-3-031-28978-1_7

constitute an important precondition for a change n the quality of surface and subsurface water. Landslides are most commonly triggered by pouring rains, in areas with an increased load on natural landscapes (motor roads, railroads, bridges and tunnels) [1–3].

For the first time, a giant landslide occurred in December 2018 at the temperature of −32 °C from the left steep shore of the Bureya reservoir. In this reach, the Bureya river crosses low mountains in the form of the erosion-denudation stretch. It is believed that the multi-stage dislacement of the landslide was preceded by intense cracking in the lower and middle parts of the slope accompanied by groundwater discharge. The displacement of the landslide led to the formation of a blockage dam that blocked off the valley of the Bureya river and the movement of water toward the Bureya hydroelectric power station [4]. To solve the problem of restoring the hydrologic regime, large-scale blasting operations were carried out. About 260 t of trotyl (trinitrotoluene – TNT) were used, and 520 sets of cumulatie shells containing hexogen were blasted.

After the avalanch of the landslide, a destructive water-ice tsunami occurred in the coastal zone of the Bureya reservoir. A high velocity of wave propagation and the physical properties of soil (high porosity, capacity for soaking, and increased plasticity) led to total destruction of soil cover to solid rock. According to the height of the tsunami wave offset, the Bureya landslide is one of the world's largest and is a unique natural phenomenon in a water body covered with ice [5]. The soil and woody vegetation on the right slope were actually destroyed (see Fig. 1). Calculaions show that about 6000 t of undecomposed organic matter of organic-soddy and peaty-organic soils, and still larger amounts of fine-grain organic matter were swept away from the area of the destroyed ground cover [5].

Upon analyzing seismic recordings, we identified threetime intervals of active deformations of the slope at the time of the landslidwe. It was concluded that destruction of the forest could occur in the first stage not only during the passage of the water wave but also as a result of the impact of the preceding "air wave" [4]. In our opinion, hypothetically, such an effect might be associated with explosive emission of methane into the atmosphere from rocks.

It was proved that as a result of a large-scale thawing of permafrost, huge reserves of carbon rapidly become an important source of methane [6]. Explosive emission of methane occurs as a result of an abrupt temperature change and might be associated with methane accumulation in the pore space of rocks at an intermittent thawing/freezing of the active layer and the thawing of permafrost [7].

The ecosystem approach to the consequences of global warming, associated with the thawing of permafrost, led to an understanding of the important role of soil microbiomes in two interrelated processes of decomposition of organic compounds and emission of greenhouse gases (CO_2 and CH_4) [8, 9]. Understanding the response of microbial communities to changes in the environment caused by warming and analysis of their functional capabilities in regions with permafrost is crucial to assessing their influence on biogeochemical cycles.

The objective of our research was to make an ecological assessment of changes in the water composition in the water area of the Bureya landslide under the influence of natural and anthropogenic factors (destruction of rocks, wash of ground cover, and use of explosives).

Fig. 1. General view of the Bureya landslide with the artificial canal (photo courtesy of A. Ostroukhov, April 2019).

2 Material and Methods

The valley of the Bureya river in the area of the landslide has a latitudinal direction. The width of the drowned channel is 500–550 m with the depth of 60–80 m. The landslide descended from the left bank about 400 m in height and 30–35° in steepness. Bedrock composing the steep left slope of the valley is represented by Proterozoic igneous rock. The fault zones show1ed an increased crevassing of rocks. The right bank is gentle, representing a terraced surface of erosion origin more than 1 km in width and up to 50 m in height above the contemporary water level. The water level fluctuations in the reservoir between maximum marks in early autumn and minimum marks in spring are 20 m [10].

According to geodetic measurements, the volume of the landslide was 24.5 mil. m^3. The volume of its portion above the water exceeds 4.5 mil. m^3, and the bulk of the landslide is under water. The parameters of the landslide are: length –800 m, heigth above sea level – from 7.5 to 46 m, and the depth of the reservoir on the site of landsslidwe descent is more than 70 m [10]. The landslide broke the ice whose thickness was 20 cm, and caused a displacement of the water mass toward the right opposite slope of the valley of the Bureya river to the height of 90 m above the water line. The water flowing up the valley of the Srednii Sandard tributary rose by 3.6 km. Most of the trees and destroyed soil cover were brought to the reservoir with the back flow of water [5].

For a comparative assessment of changes in the water quality in the Bureya reservoir, water samples were cllected around the landslide body and from the artificial canal (January–April 2019). In the summer period of 2020, additional meltwater samples were collected from coastal aufeis, and water samples from the Pervyi and Srednii Sandar Gulfs which were filled with floating wood (see Fig. 2).

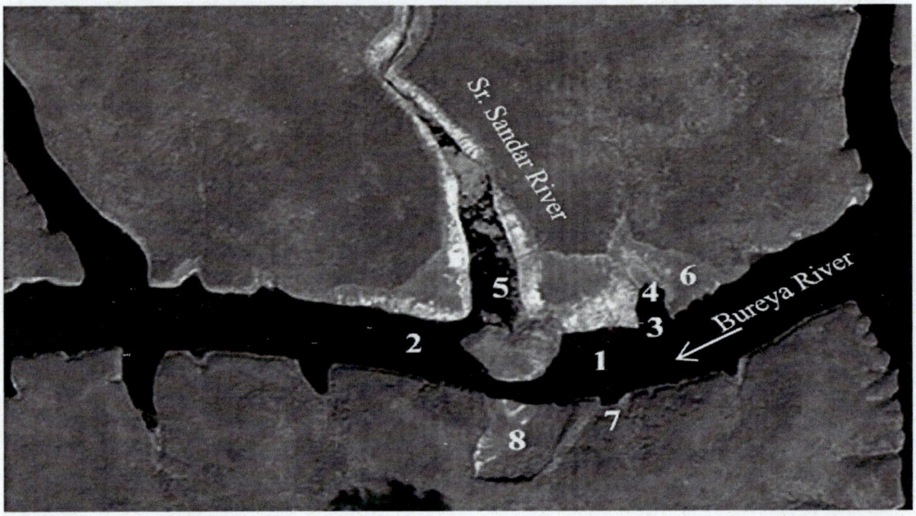

Fig. 2. Satellite image of the landslide zone (Sentinel-2A, 7.6.2019) and of the water sampling location: 1 – upstream of the landslide body, 2 – downstream of the landslide body, 3 – Pervyi Gulf, 4 – Pervyi brook, 5 – Srednii Sandar Gulf, 6 – Golubaya Naled brook, 7 – aufeis, 8 – site of the landslide.

To remove fine-grained suspended solid material and determine the elemental composition, water samples were prefiltered through a filter (pores 2–3 μm in dwiameter) and preserved with freshly prepared nitric acid (1:1). In 50 mL test tubes, we added 3 mL of HNO_3 solution and 100 μL of nitrate solution In with a concentration of 20 ppm, as the internal standard. The content of the elements in the water was determined in the Innovation Analytical Center (at ITG FEB RAS) by inductively coupled plasma mass spectrometry with ICP-MS ELAN-9000 (Perkin Elmer, USA) (as per GOST R 56219-2014; ISO 17294-2:2003). The determination of the composition of volatile organic compounds (VOCs) in water samples was carried out in the Ecological Monitoring Center at IWEP FEB RAS by the method of gas chromatography using HP-FFAP (50 m; 0.320 mm; 0.50 μ) at the temperature interval 50–230 °C (MUK 4.1.3171— 14). To estimate the contribution of microbial complexes (MC) to the transformation of OM in the water bodies under investigation, we used the population of various physiological groups of bacteria: CHB – cultivated heterotrophic bacteria, AMB – ammonifying bacteria, NB – nitrifying bacteria, HTB – humate transforming bacteria, PRB – phenol-resistant bacteria, SRB – sulfate-reducing bacteria. Cultivation was carried out on agarized culture media with different sources of carbon (peptone, starch, Na humate, and Ca lactate), in accordance with conventional methods of microbiology. The abundance of microorganisms was expressed in terms of colony forming units per 1 mL of water (CFU/mL).

3 Results and Discussion

3.1 Comprehensive Assessment of the State of the Water Environment in the Winter Period

The investigations that were made in the winter period involved assessing changes in the water quality as a result of the descent of a he mass of landslide into the reservoir and using explosives for restoring the hydrologic regime and renewing access of water to the Bureya hydroelectric power station. After the landslide, the natural waters of the reservoir received OM (organic matter) of different structure which were contained in the surface soil layer and in the pore space of crushed rocks.

Spectrophotometry was used to obtain an increase in total content of organic matter ($OM_{254)}$ and aromatic compounds ($OM_{275)}$ downstream of the landslide body as well as in the canal immediately after the blasting operations in February 2019, while OM content began to decrease in March (Table 1). A change in the water quality in the Bureya reservoir after the blasting operations could, to a larger extent, be promoted by VOCs of a different genesis: products of TNT detonaton and decomposition arriving from the pore space of crushed rocks. It was established that in natural conditions there is taking place an aerobic and anaerobic transformation of TNT which is able to give rise to the more toxic aminoaromatic products of bacterial metabolism [11].

Strangely enough, but such a large-scal process of destruction of the base of the slope and input of rocks into the reservoir might be associated with microbiological processes triggered by intermittent thawing/freezing of rocks saturated with water. Analysis of the composition of VOCs (Table 1) revealed that many of them can be assigned to products of microbial metabolism which are produced as a result of the methanogenesis and methanotrophy. It is known that alkanes (methane, ethane, propane, and hexane) are present in sedimentary rock, coal seams, and bog water. As a result of their enzymatic oxidation/hydrolysis, diverse metabolites, including acetone, butanol, etc., are produced [12]. The genesis of methanol that is present in many samples migrht be attributed to the transformation of natural methane [13]. We suggested the hypothesis of the cause of the Bureya landslide, namely methane emission due to an abrupt decrease in temperature [14].

Sensitive bioindicators of input of OM of different structure into the water were represented by cultivated heterotrophic bacteria (CHB). An increase in the population of CHB downstream of the landslide body was due to slow drainage of water through loose material and washing of OM from the pore space. The dynamics of the CHB population reflected diversity of the OM composition in water samples. Their maximum abundance was observed in the flow canal upon completion of the blasting operations (February 2019).

Table 1. Comprehensive assessment of the water quality in the Bureya reservoir in the area of the landslide during the winter-spring period of 2019.

Sampling location		OM content			Number of CHB, CFU/mL
		OM_{254}	OM_{275}	Composition of VOCs	
Prior to blasting operations, 25.1.2019					
Above the landslide	LB	0.324	0.282	Acethaldehyde, acetone, methanol, benzene, tolouene	111 ± 9
	RB	0.303	0.196	Acethaldehyde, acetone, methanol, *m*-xylene, isopropylbenzene	101 ± 3
Below the landslide	LB	0.524	0.312	Acetone, methanol, benzene, toluene, butyl acetate, ethylbenzene, xylenes, isopropylbenzene	63 ± 2
	RB	0.587	0.324	Hexane, acetone, methanol, benzene, toluene, butyl acetate, *o*-, *m*-xylenes, butanol, isopropylbenzene	144 ± 10
After blasting operations					
Water samples from the canal (14.2.2019)		0.682	0.494	Acethaldehyde, acetone, methanol, toluene, *o*-xylene, isopropylbenzene	315 ± 3
		0.521	0.321	Acethaldehyde, acetone, toluene, *o*-, *m*-xylenes	173 ± 8
		0.753	0.545	Hexane, acethaldehyde, acetone, ethyl acetate, methanol, tolouene, *o*-, *m*-xylenes, isopropylbenzene	484 ± 12
		0.724	0.514	Hexane, acethaldehyde, acetone, methanol, toluene, *m*-xylene, isopropylbenzene	414 ± 3
Water samples from the canal (1.3.2019)		0.385	0.211	Hexane, acethaldehyde, acetone, ethyl acetate, methanol, toluene, *o*-xylene, isopropylbenzene	133 ± 12

Note: LB, RB – left bank, right bank.

3.2 Microbiological Investigations in the Summer Period of 2020

An important ecological factor determining the water quality in the Bureya reservoir in the area of the landslide in summer 2020 was represented by floating and drowned wood in the gulfs along the right bank (see Fig. 3A, 3B). Taking into consideration that wood undergoes microbial decomposition to produce various transformation products, microbiological investigations were made upstream and downstream of the landslide body as well as in the nearby gulfs filled with wood.

The highest number of many physiological groups of bacteria participating in the decomposition of vegetable debris was observed in the Srednii Sandar Gulf (Table 2). The summer season of 2020 was characterized by a low water level, as a result of which the gulf was isolated from input of water masses from the reservoir, and water samples were collected from zones with a weak water exchange. A high level of color of the water in the gulf is associated with its increased humification as a consequence of the transformation of lignocellulose to produce oxidated forms of aromatic compounds, including humic substances and benzoquinones.

Fig. 3. Accumulations of wood in the water area of the Bureya reservoir and coastal aufeis: A – Pervyi Gulf; B – Srednii Sandar Gulf; C – Golubaya Naled brook on the right bank; D – aufeis on the left bank (photo courtesy of A. Ostroukhov, July 2020).

A high activity of the microbial complex (MC) participating in the decomposition of humic substances is proved by a high abundance of bacteria utilizing Na humate *in vitro*. The st ructure of MC showed a high share of PRB whichare indicators of phenol

Table 2. Abundance of different physiological groups of bacteria in the water area of the Bureya reservoir in the zone of influence of the landslide (July 2020).

No.	Sampling location	CHB	AMB	HБ	HTB	PRB	SRB
1	Above landslide body	301 ± 9	256 ± 31	277 ± 45	273 ± 66	9 ± 4	50 ± 3
2	Below landslide body	82 ± 37	98 ± 56	9 ± 2	37 ± 9	0	107 ± 11
3	Pervyi Gulf*	584 ± 36	814 ± 1	680 ± 18	497 ± 14	13 ± 4	191 ± 15
4	Pervyi brook	181 ± 10	277 ± 10	120 ± 11	168 ± 10	9 ± 4	190 ± 20
5	Srednii Sandar Gulf	710 ± 42	1600 ± 61	693 ± 12	1717 ± 109	550 ± 6	510 ± 22
6	Golubaya Naled brook	501 ± 57	225 ± 28	407 ± 66	310 ± 16	239 ± 99	244 ± 38
7	Aufeis	537 ± 18	143 ± 16	377 ± 9	480 ± 7	0	953 ± 17

Note: * – amidst floating wood; CHB – cultivated heterotrophic bacteria, AMB – ammonifying bacteria, NB – nitrifying bacteria, HTB – humate transforming bacteria, PRB – phenol-resistant bacteria, SRB – sulfate-reducing bacteria.

pollution of the aquatic environment. A similar microbiological situation was observed in the Pervyi Gulf amidst floating wood. In the area of the gulf where there was taking place the water exchange with the reservoir the number of all physiological groups of bacteria wasconsiderably lower.

Winter groundwater seeps that form aufeis can serve as the sources for OM input into the reservoir. In the summer period of 2020, two aufeis formations were investigated. One of them melted and flowed down in the form of the Golubaya Naled brook (see Fig. 3C). It is known that the OM composition in the natural ice is formed under the influence of cryochemical and biochemical processes. The aufeis formations showed the presence of saturated and unsaturated hydrocarbons, methane, carbon oxides, ammonia, and hydrogen sulfide [15]. The composition of groundwater interacting with rocks and soils is derived from contents of different components in aufeis. The microbial community of the Golubaya Naled brook (Table 2) was characterized by rich variety and a large populat4ion of most physiological groups of bacteria whose development was ensured by transformation products of vegetable and terrigenous material arriving from the disturbed soil cover. A considerable proportion can be made up by phenol compounds, and this is confirmed by a high abundance of PRB and bacteria growing on Na humate. Another aufeis formation (see Fig. 3D), located on the left bank 3 km from the site of the landslide during the expedition did not melt completely. Bacteria utilizing a broad spectrum of OM, including humic substances, were present in the melt water arriving from beneath a huge block of ice.

It is worth paying attention to the group of sulfate-reducing bacteria which are able to utilize plant remains when oxygen is lacking. They were present in large amounts in water samples collected amidst wood in stagnant zones as well as in aufeis. They play an important role in migration of a number of elements and are producers of hydrogen sulfide.

3.3 Dynamics of the Elemental Composition of Water After the Landslide

In the winter period, the main factors influencing the content of lithogenic elements in the water in the zone of influence of the landslide were represented by crushing of rocks, drainage of water through the body of the landslide, and blasting operations. Downstream of the body of the landslide at either bank, an increase in concentration of lithogenic elements Fe, Al and Pb could occur due to drainage of composing rocks, the interaction of water with minerals, and their transformation into a soluble state (Table 3). On the contrary, the Ca, Mg, Zn and Cd concentrations were decreasing, possibly due to their sorption in the porous space of rocks.

The presence of OM in the aquatic environment is an important factor governing the sorbtion-desorbtion processes of many metal cations. For instance, the behavior of Ni and Cu depends on the production of organometallic complexes, while the trivalent Fe, entering into the reaction with humic acids, transforms into a soluble state [16]. The

Table 3. Elemental composition of water in the zone of influence of the Bureya landslide.

Elements	January 2019				July 2020			
	LB, above the landslide	LB, below the landslide	RB, above the landslide	RB, below the landslide	M, above the landslide	M, below the landslide	Pervyi Gulf	Srednii Sandar Gulf
Mg	1468.0	1350.7	1457.8	1375.5	62.3	73.6	135.4	158.9
Al	164.0	181.8	165.8	185.3	36.2	9.7	53.0	15.4
Ca	4102.5	3874.4	4160.2	3907.8	136.7	59.3	186.1	199.3
Cr	1.0	1.2	1.0	1.4	0.4	0.7	0.4	0.4
Mn	11.5	9.9	7.7	9.5	4.2	2.0	11.0	3.6
Fe	207.9	344.1	274.2	338.4	5.0	39.3	79.8	bd
Co	0.1	0.1	0.1	0.1	0.1	2.8	1.0	4.8
Ni	1.1	0.8	0.7	2.9	0.4	0.8	0.7	0.7
Cu	5.6	5.7	4.0	5.1	2.7	6.1	2.9	4.1
Zn	22.4	17.0	14.2	11.3	22.5	29.1	33.1	23.6
As	0.5	0.5	0.5	0.5	Bd	Bd	0.2	Bd
Cd	0.1	bd	0.2	bd	Bd	0.1	0.1	0.1
Pb	0.3	1.3	bd	1.4	0.3	1.9	1.7	6.3

Note: LB – left bank, RB – right bank, M – middle, bd - below detection.

most dramatic changes in contents of many elements in the water were observed after the blasting operations. They led to a decrease in Mg, Al and Ca concentrations in the artificial canal as compared to their content in the water collected around the landslide body [17]. A special phenomenon was associated with an abrupt decrease in concentrations of elements, such as Fe, Cu, Zn and Pb. The effect of decrease in concentrations of these elements can be attributed to the various biogeochemical processes on the surface of rocks crushed by explosions. The oxidized forms of Fe are able to sorb a number of heavy metals in the presence of OM. Cu, Zn and Pb possess the largest affinity to the oxide forms of Fe [18]. Based on this, it becomes clear why the concentrations of Fe, Cu, Zn и Pb decreased simultaneously.

In the summer period of 2020 there occurred an abrupt decrease in concentration of many elements in the water of the reservoir in the area of the landslide (Table 3). In water samples collected from the Srednii Sandar Gulf, amidst floating wood we observed a decrease in Fe content to values below the detection limits ofe the instrument. An important role could be played by OM which was brought in with soils and wood debris. It is known that wood is a multilayered porous material containing high-molecular structures [19]. Functional groups of lignin (phenol, methoxy- and hydroxyl) can interact with metal ions. Cellulose and hemicellulose contain a large amount of polar groups which adsorb ions of heavy metals quite well [20].

4 Conclusion

The Bureya landslide became the cause for prolonged ecological risks which are associated with input of crushed rocks, soil humus and wood debris into the water with the overland runoff during the spring snowmelt and summer floods. The biochemical transformation of OM of a different structure in the water and bottom sediments leads to an increase in content of soluble methylated benzene derivatives and toxic elements and t4o an increase in water color. An intensification of the processes of microbiological decomposition of wood debris with the involvement of sulfate-reducing bacteria can be accompanied by the formation of hydrogen sulfide zones in shallow gulfs thus posing a threat to hydrobionts of the reservoir.

A poor knowledge of the process of explosive methane emission in regionswith retreating permafrost can lead to the risk of increasing negative geological events in the winter period within the Bureya river basin, including the appearance of craters near coal deposits, destruction of the infrastructure of mines and hydrocarbon storage facilities or future landslides on the steep shores of the Bureya reservoir.

References

1. Badoux, A., Andres, N., Techel, F., Hegg, C.: Natural hazard fatalities in Switzerland from 1946 to 2015. Natural Hazards and Earth System Science **16**(12), 2747–2768 (2016)
2. Andres, N., Badoux, A.: The Swiss flood and landslide damage database: normalisation and trends. J. Flood Risk Manag. **12**(S1), e12510 (2019). https://doi.org/10.1111/jfr3.12510
3. Zheng, Q., Shen, S.-L., Zhou, A.-N., Cai, H.: Investigation of landslides that occurred in August on the Chengdu-Kunming Railway, Sichuan, China. Geosciences **9**(12), e497 (2019)

4. Zerkal, O.V., Mahinov, A.N., Kudymov, A.V., Haritonov, M.E., Fomenko, I.K., Barykina, O.S.: Bureinskiy landslide 11 of December 2018. Formation condition and and features of development mechanism. Georisk **8**(4), 18–30 (2019)
5. Mahinov, A.N., Mahinova, A.F., Levshina, S.I.: Assessment of the flushes of the soil cover by the water-ice tsunami and the water quality in the area of the landslide at the Bureyskoe Reservoir. Meteorology and Hydrology **11**, 64–73 (2020)
6. Singleton, C.M., et al.: Methanotrophy across a natural permafrost thaw environment. ISME J. **12**, 2544–2558 (2018)
7. Buldovicz, S.N., et al.: Cryovolcanism on the Earth: origin of a spectacular crater in the Yamal Peninsula (Russia). Scientific Reports V **8**, e13534 (2018)
8. Kraev, G.N., Rivkina, E.M.: Accumulation of methane in freezing and frozen soils of the permafrost zone. Arctic Environmental Research V **17**(3), 173–184 (2017)
9. Kwon, M.J., Jung, J.Y., Tripathi, B.M., Göckede, M., Lee, Y.K., Kim, M.J.: Dynamics of microbial communities and CO2 and CH4 fluxes in the tundra ecosystems of the changing Arctic. Microbiol. **57**(5), 325–336 (2019)
10. Mahinov, A.N., Kim, V.I., Ostrouhov, A.V., Matveenko, D.V.: A large landslide in the Bureya river valley and a tsunami in the reservoir of the Bureya hydroelectric power station. Vestnik DVO RAN **2**, 35–44 (2019)
11. Kalderis, D., Juhasz, A.L., Boopathy, R., Comfort, S.: Soils contaminated with explosives: environmental fate and evaluation of state-of-the-art remediation processes. Pure Appl. Chem. **83**, 1407–1484 (2011)
12. Meslé, M., Dromart, G., Oger, P.: Microbial methanogenesis in subsurface oil and coal. Res. Microbiol. **164**(9), 959–972 (2013)
13. Yu, Z., Beck, D.A., Chistoserdova, L.: Natural Selection in Synthetic communities highlights the roles of methylococcaceae and methylophilaceae and suggests differential roles for alternative methanol dehydrogenases in methane consumption. Front. Microbiol. **5**(8), e2392 (2017)
14. Kondratyeva, L.M., Litvinenko, Z.N., Filippova, G.M.: Environmental risk of VOC formation after a major landslide. Geoecology **3**, 74–81 (2020)
15. Ivanov, A.V.: Cryogenic Metamorphization of the Chemical Composition of Natural Ice, Freezing and Melt Waters. Dalnauka, Vladivostok, p. 164 (1998)
16. Wan, X., Xiang, W., Wan, N., Yan, S., Bao, Z., Wang, Y.: Complexation and reduction of iron by phenolic substances: implications for transport of dissolved Fe from peatlands to aquatic ecosystems and global iron cycling. Chem. Geol. **498**, 128–138 (2018)
17. Kondratyeva, L.M., Golubeva, E.M., Litvinenko, Z.N.: Biogeochemical assessment of changes in water composition after a major landslide in winter. Pacific Geography **2**, 43–53 (2021)
18. Perelomov, L.V., Pinsky, D.L., Violante, A.: Influence of organic acids on the adsorption of copper, lead and zinc by goethite. Pochvovedenie **1**, 26–33 (2011)
19. Mi, R., et al.: Scalable aesthetic transparent wood for energy efficient buildings. Nature Communications V **11**, 3836 (2020)
20. Sun, S.C., et al.: Effect of integrated treatment on improving the enzymatic digestibility of poplar and the structural features of isolated hemicelluloses. Carbohydr. Polym. **252**, 117164 (2021)

Floro-Cenotic Features of Vegetation in the High Mountains of Northwestern Cisbaikalia

Elena Korolkova(✉)

V.B. Sochava Institute of Geography, Siberian Branch, Russian Academy of Sciences, 1, ul. Ulan-Batorskaya, 664033 Irkutsk, Russia
elainefisher@yandex.ru

Abstract. Long-term investigations of vegetation in the high-mountains of North-eastern Cisbaikalia within the boundaries of the Baikal Mountains were carried out. Methods of geobotanical mapping were used to identify the structural organization of vegetation communities and their complexes within the goletz and subgoletz belts. Their distribution patterns are illustrated in a cartographic model at a scale of 1:500 000. Cenotic diversity of vegetation communities is represented in the map legend that was compiled on the principles of a multidimensional and multistep classification of vegetation as developed by V.B. Sochava. The legend includes six cenotic subdivisions integrating plant associations and their groups into complexes and combinatins. Each name in the legend number provdes Latin names of dominant species of plants and indicator plants listed the communities in decreasing order. A total of two types of vegetation communities of high mountains were identified: mountain-taiga communities, and communities of mountain tundras. To illustrate the smoothness of the transition of high-mounain vegetation to taiga vegetation the map also shows three epitaxa of taiga type occurring in the subgoletz belt.

Keywords: Northeastern Cisbaikalia · geobotanical mapping · classification of vegetation · phytocenotic diversity

1 Introduction

Northwestern Cisbaikalia is a complex mountain-depression system combining various factors which influence the structre and development of vegetation. The territory is incorporated into the zone of interaction of several centers of floristic and phytocenogenetic formation at t4he intrface of physical-geographical regions of North Asia: the Baikal-Dzhugdzhur, Middle-Siberian and South-Siberian regions [1] which shaped themselves in the process of complex evolutionary-genetic transformations of the geosystems. Their interpenetration promotes emergence of a great variety of vegetation communities in the region. An enormous contribution to the study of vegetation in this hard-to-reach region was made by botany and floristics scientists M.G. Popov, L.N. Tyulina, A.N. Lukicheva, L.I. Malyshev, G.A. Peshkova, A.V. Belov, V.N. Molozhnikov, L.V. Bardunov, as well as by N.V. Stepantseva, T.V. Mokryi, S.G. Kakzanovskii, and V.V. Chepinoga [2]. I.N. Vladimirov caried out a number of studies of forest vegetation [3]. In summarizing

I. N. Vladimirov et al. (Eds.): RERSDNA 2022, SPEES, pp. 68–76, 2023.
https://doi.org/10.1007/978-3-031-28978-1_8

archival and full-scale observational data for the territory of Northwestern Cisbaikalia, the investigations were continued by using the method of geobotanical mapping [4]. The main objectives of this paper are to identify the general geographical-genetic trends in vegetation, analyze the structure and visualize the principal distribution patterns of vegetation communities.

2 Material and Methods

From available data sources it became clear that the natural-climatic characteristics of the study area are quite varied. According to the regionalization of the south of North Asia, Northwestern Cisbaikalia is assigned to the Baikal-Dzhugdzhur physical-geographical region [1]. The territory includes the mountain ranges: Baikal Mountains, Anai goltsy, and southwestern part of the Ungdar Ridge. The morphological features of topography permit it to be assigned to the high-mountain type with pronounced landforms and vertical dissection, high uplands and deep intermontane valleys and depessions. The mountain ranges are composed largely of Proterozoic and Cambrian formations with built-in unevenly-aged intrusive rocks [5].

According to the existing soil regionalization, the study territory forms part of the province of podzols, podburs and soddy forest soils of the mountains in Cisbaikalia and Stanovoi Highlands as well as of the province of podzolic, soddy forest, soddy calcareous and gray forest soils of the Irkutsk amphitheater [6].

The climatic characteristics of the region depend on its location in the center of Asia where the Asian anticyclone is dominant as well as on topography that forms the mountain barrier-shadow and depressin effects [7]. The transport of northwesterly air masses is dominant over the territory. This is the cause for the asymmetric climatic situation of the western and eastern macroslopes of the mountain ridges. On the whole, the climate of Cisbaikalia is defined as moderately continental.

Expedition-based data provided a groundwork for identifying and assessing the diversity of vegetation and its present status. The investigations were made in three stages. In the period from 2004 to 2008, soil, herbarium and peat samples were collected for palynological investigation, and wood cores were taken for dendro-chronological datings and refinements of age changes in geosystems and their interrelationship with climate dynamics. A generalized map of "Modern vegetation of Northwestern Cisbaikalia" at a scale of 1:500 000 was compiled [2]. After that, in 2011 and during 2015–2021, extrapolated data were verified for key profiles and mappping units for the entire study territory along the high-mountain belt of Northwestern Cisbaikalia, and the assessment of the pyrogenic and anthropogenic disturbance of vegetation communities was updatedby using interpretation of Landsat 7 and Sentinel 2 imagery. A combination of the remote sensng and ground-based methods of studying the secular vegetation dynamics is necessary for real-time updating of cartographic data [8, 9]. The studies were all made on the basis of the QGIS 3.22.11 software package.

Primary vegetation communities in the spectrum of centuries-long changes in vegetation were identified by using the method of palynological investigations of peat samples collcted on the watershed divide of the Baikal Mountains and in their foothills [10, 11].

A cartographic study and classification of vegetaion cenoses can be performed by using different methods and approaches [12]. This paper uses, as the basic approach in the classification of vegetation, the principle of multidimensional classificaion of vegetation communities as suggested by Academician V.B. Sochava [13] and improved by his followers [14]. The main principle in this method of mapping involved displaying not only modern vegetation of the territory but also its secular dynamics and evolutionary changes. Determinations of the plant species were performed in accordance with the generally accepted systematics; the usage of Latn names has been clarified [15].

3 Results and Discussion

The map entitled "Modern vegetation of the high mountains of Northwestern Cisbaikalia" was compiled using material collected (see Fig. 1). The figure shows a fragment of the map, and the table presents a complete list of diversity of high-mountain vegetation cenoses for the entire territory.

The findings of the investigation showed that vegetation in the high mountains of Northwestern Cisbaikalia is composed of combinations of communities of the Altai-Tian-Shan Alpine and Mountain-Tundra phratries of vegetation formations. A clear-cut determination of the boundaries of the goletz and subgoletz belts in this mountainous region is difficult. The smoothing effect of the mesoclimate of Baikal's water area, a high continentality, and also the topographic features of the region (the altitude of the mountain ridges averages 1900 m above the sea level) were responsible for the active interpenetration of representatives of the high-mountain flora into the mountain-taiga belt, and vise versa. The investigations showed that the main regularity in the succession of vegetation communities is typified by a gradual change in the general climatic indicators during the meridianal movement. In the southern part of the territory, vegetation more actively indicates a succession of underlying rocks. Let us now consider in greater detail the changes in the composition of high-mountain groups of associations from south to north along the axis of the watershed divides in Northwestern Cisbaikalia.

In the south of the territory, along a gentle watershed divide of the Baikal Mountains from the source of the Rytyi brook to the upper reaches of the brooks of the Bol. Solontsovyi Cape there occurs the mesocomplex of the communities of the South-Siberian vegetation formation: forbs-shallow turf-grass steppizated and meadowed tundras in combination with barren meadows [16]. The barren meadows are characterizedby a combination of plant species, such as *Helictotrichon mongolicum, Festuca lenensis, F. ovina, Hierochloe alpina, Carex ledebouriana, C. ensifolia, Luzula sibirica, Salix berberifolia, Phlojodicarpus villosus, Dryas oxyodonta, Sibbaldia procumbens, Potentilla nivea, Anemonastrum sibiricum, Pedicularis amoena, Patrinia rupestris, Aconogonon ajanense, Minuartia arctica,* with inclusion of *Sorbaria pallasii*, etc. The communities are sparse due to outcrops of underlying rocks with lithophilic vegetation (*Orostachys spinosa, Thymus sp., Campanula rotundifolia subsp. Langsdorffiana, Selaginella rupestris, Youngia tenuifolia*) and screes with crustaceous lichens. The grass layer is virtually absent. In the same area but along the macroslope of the Baikal Mounains, near the sites of thawing of snow patches, there occur communities of the Baikal-Dzhygdzhyr vegetation formation: sparse thickets of dwarf birch (*Betula exilis*) as well as wasteland of subshrubs: foxberries, crowberries, bog ledum and subshrub willow (*Salix sphenophylla*). Not rich forbs were observed in them, consisting of *Carex williamsii, Luzula sibirica, Anemonastrum sibiricum, Dasystephana algida, Hierochloe alpina,* and *Aconogonon ajanense.* Sparse thickets of dwarfish Siberian mountain pine occur along the entire goletz belt concentrating on large screes and rock outcrops. These groups of associations grow on acidic rocks dating back to the Lower Proterozoic.

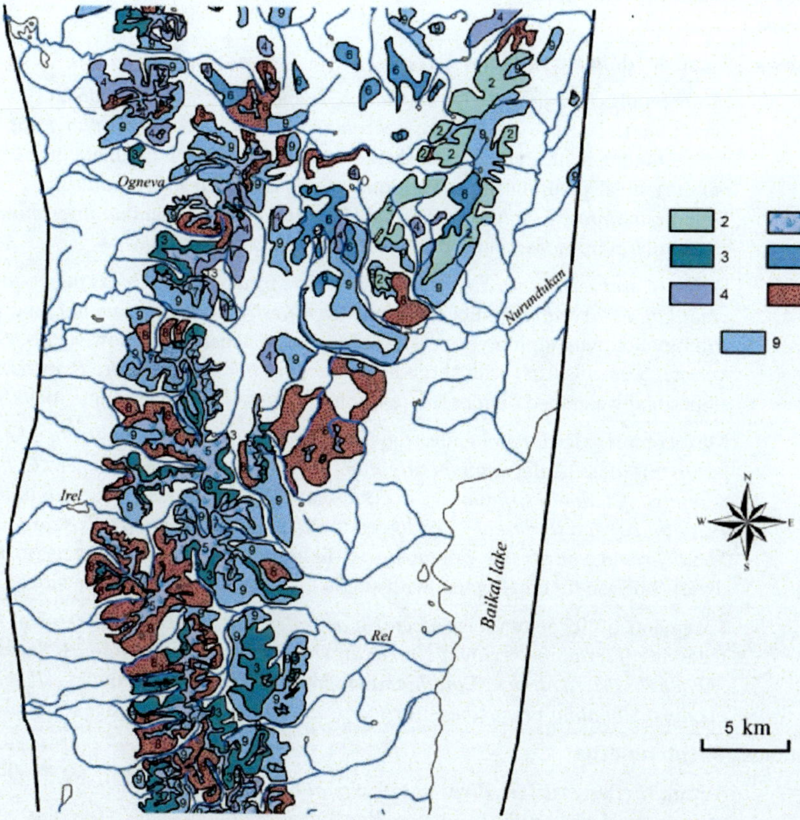

Fig. 1. A fragment of the 1:500 000 map "Modern vegetation of the high mountains of Northwestern Cisbaikalia". For the map legend, see the Table 1.

North of the Bol. Solontsovyi Cape as far as the Sev. Kedrovyi Cape along the eastern macroslope and gentle watershed divides there is taking place a substitution of steppizated and meadowed tundras with barren subshrub-sparse grass and lichen "cushion-shaped" tundras forminga mosaic pattern in accordance with the forms of micro- and mesorelief.

The high-mountain plateaus (1700–1950 m) still retain elements of the steppe flora. In rocky tracts with alpine peaked forms and in depressions between them (1850–2100 m) there occur bluegrass-subshrub communities poor in diversity consisting of bilberries and bog ledum, with sparce dwarf Siberian mountain pine.

Table 1. Legend to the map "Modern vegetation of the high mountains of Northwestern Cisbaikalia" (nubers stand for forest unit numbers in the legend).

Type of vegetation:	**High-mountain vegetation**
Class of formations:	**South-Siberian formations**
Formations:	**Meadowed lichen tundras**
1	Mesocomplex: cetraria-cladonia (*Cladonia alpestris, Cetraria nivalis, C. cucullata, C. islandica, Alectoria ochroleuca*) с дриадой (*Dryas punctata*) rubble tundras with fragments of thickets of dwarf Siberian mounain pine (*Pinus pumila*), nival small meadows (*Trollius asiaticus, Aquilegia glandulosa, Saussurea alpina, S. subacaulis, Calathiana uniflora, Campanula turczaninovii, Equisetum scirpoides*) and screes on slopes and gentle surfaces
2	Sedge (*Carex tristiss sp. Stenocarpa, C. podocarpa*) -dryas (*Dryas punctata, D. crenulata,D. oxyodonta*) -lichen (*Cladonia alpestris, Cetraria nivalis*) tundras in combination with alpinotype small meadows and separated willow stands (*Salix krylovii,S. taraikensis*) with shrub birch trees (*Betula nana. Subsp. B. exilis*) on slopes and watershed divides, more often with outcrops of carbonate rocks
3	Mesocomplex: *R. aureum* -subshrub (*Vaccinium vitis-idaea, V. myrtillus, Ledum palustre*) -moss (*Aulacomnium turgidum, Rhytidium rugosum*) –lichen (*Cetraria islandica, Cladonia sp.*) tundras in combinations with grasses (*Festuca altaica, Poa altaica*) and *R. aureum* meadowed tundras, sedge (*Carex tristis subsp. Stenocarpa, C. atrofusca, C. globularis*)-lichen communities and tracts of sparse dwarf Siberian mountain pine, with single birch trees, most often on slopes
4	Fragments of lichen tundras (*Alectoria ochroleuca, Cetraria nivalis*), nival small meadows (*Geranium krylovii, Oxyria digyna, Ranunculus sulphureus, Saussurea subacaulis, Viola altaica, Calathiana uniflora*) and meadowed tundras (*Potentilla elegans, Sibbaldia procumbens, Cassiope ericoides*) on rocky edges and taluses
Formations:	**Shrub tundras**
5	Yernik (*Betula exilis, B. divaricata*) subshrub (*Vaccinium myrtillus, V. vitis-idaea, Ledum palustre*)-moss (*Racomitrium lanuginosum, Grimmia ovata*)-lichen (*Cetraria islandica, Cladonia mitis, C. rangiferina*) mountain tundras, in places subshrub-lichen-moss, with separated stands of dwarf Siberian mountain pine and singly occurring larch trees (*Larix sibirica, L. dahurica*) on gentle planation surfaces
Fomations:	**Subalpinotype meadows**
6	Subalpinotype meadows (*Aquilegia glandulosa, Saussurea alpina, S. parviflora subsp. Purpurata, Rhodiola rosea, Cirsium heterophyllum, Bistorta attenuata, Trollius asiaticus, Delphinim elatum, Saxifraga nelsoniana, Valeriana altaica*) in combination with alpinotype meadows in the2 vicinty of snow patches and small brooks and in combinationwith meadowed tundras and thickets of shrubs (*Pinus pumila, Salix krylovii, Betula exilis*)

(*continued*)

Table 1. (*continued*)

Type of vegetation:	**High-mountain vegetation**
Type of vegetation:	**Taiga (boreal) vegetation**
Class of formations:	**Baikal-Dzhugdzhur formations**
Formations:	**Subgoletz vegetation of dwarf Siberian mountain pine**
7	Sparse thickets of dwarf Siberia mountain pine subshrub (*Vaccinium myrtillus, V. vitis-idaea*) -lichen (*Cladonia amaurocraea, Cl. Rangiferina*) in combination with fragents of mountain tundras, nival meadows and screes
8	Closed thickets of dwarf Siberian mountain pine with R. aureum in combination with singly occurring larch trees (*Larix sibirica*) and Siberian stone pine (*Pinus sibirica*) subshrub (*Ledum palustre, Empetrum nigrum*) -lichen (*Cladonia amaurocraea, Cl. Rangiferina, Cetraria nivalis, C. cucullata*) in combination with mountain tundras and screes
9	Dwarf Siberian mountain pine-yernik (*Betula rotundifolia*) and yernik-dwarf iberian mountain pine comunities subshrub (*Vaccinium vitis-idaea, Arctostaphylos uva-ursi, Ledum palustre, Empetrum nigrum*) -forbs (*Geranium eriostemon, Dendranthema zawadskii, Lathyrus humilis, Galium boreale, Artemisia tanacetifolia*) -moss (*Polytrichum hyperboreum, Rhytidium rugosum*) -lichen (*Cetraria nivalis, C. cucullata, C. islandica, Cladonia amaurocraea*) in combination with rarely occurring larch (*Larix sibirica*) and Siberian stone pine

From the Sev. Kedrovyi Cape to the Elokhin Cape, with a decrease in the height of the ridge to 1700 m abve the sea level, goletz vegetation is dominated by groups of associations of the Baikal-Dzhugdzhur formation: mountain-tundra communities with sparse dwarf Siberian mountain pine and *Rhododendron aureum* and relatively poor mosaic subshrub-sparse grass inclusions in combination with stony-rubble areas consisting of gray and green porphyries.

From the Elokhin brook and just north of the Khibilen Cape, the underlying roks in the goletzes are substituted by carbonate and weakly carbonate rocks. There occur on them dryas-subshrub and subshrub-lichen mountain tundras of continuous coverage. Such meadows usually occur on sites of thawing of snow patches, the shores of high-mountain lakes and sources of brooks. Shortgrass alpinotype and tall-grass subalpinotype meadows are represented by South-Siberian formations [4] are occur on the territory relatively rarely: the northeastern boundary of their occurrence runs in Cisbaikalia.

In the center of the Baikal Mountains, from the Khibilen Cape to the B. Kosa Cape, the goletzes show a sparse-cover occurrence of the Baikal-Dzhugdzhur mounain tundras consisting of subshrub-sparse grass-lichen communities in combination with sparse thickets of dwarf Siberian mountain pine [17]. A characteristic change of mountain-tundra vegetation is associated with an increase in the heigh of the mountain ridge to 1950–2200 m above the sea level as well as with the changed composition of underlying rocks again from carbonate to acidic and weakly carbonate intrusive [18]. Another factor for the succession of vegetation involved changes in mesoclimatc conditions in the movement from south to north. In the areas of the Molokon, Muzhinai and Kurkula rivers, the mountain massif becomes broader (asbout 40 km at the base), nd the main watershed ridge moves away 8–10 km from Baikal's shore. In this area, the mountain ridge reaches its maximum height: Mount Chersky 2588 m above the sea level. The mountain ridge has pointed, deply horizontally dissected peaks. There appear alectoria mountain tundras in combinat4ion with rock outcrops, sometimes covered with epilithic lichens of the species: *Rhizocarpon geographicum, Xanthoria elegans, Lecidea advertens.* More often here occur alpinotype and subalpinotype meadow groups of associations of t4he abundant forbs: *Aquilegia glandulosa, Anemonastrum sibiricum, Ranunculus propinquus, Anthoxanthum alpinum, Viola altaica, Calathiana uniflora, Carex podocarpa, C. media, C. magellanica, Bistorta vivipara, Veratrum lobelianum, Allium schoenoprasum, Oxyria digyna, Trollius kytmanovii, Rhodiola rosea.*

In moving northward, the height of the Baikal mountains decreases; however, there emain peaked alpinotype forms of high mountains. A lowering of the boundary of penetration of shrub tundras into the goletz belt is observed. The Baikal Mountains smoothly turn their direction north-eastward, moving away 15–25 km from Baikal and expanding up to 45–50 km at the foothills. Across the entire territory of the high-mountain belt there occur South-Siberian formations of cetraria and alectoria mountain-tundra groups of associations on basic and alkaline underlying rocks. In flattened areas they are accompanied by groups of associations of the Baikal-Dzhugdzhur formations of subshrub and lichen-subshrub mounain tundras in combination with mosaic thickets of dwarf Siberian mountain pine and *R. aureum.* In humid depressions there occur sedge (*Carex bigelowii subsp. Ensifolia, C. podocarpa, C. saxatilis*) –moss (*Aulacomnium turgidum, Dicranum elongatum*) communities with subshrubs (*Vaccinium uliginosum, Ledum palustre*). Large areas along brooks and boggy tracts are occupied by shrubs consisting of willows *Salix saxatilis, S. polaris* and yernik birch trees *Betulala nata, B. divaricata, B. exilis.*

Futher out to the north, along the mountain ridges from the Davanskii pass to the Okunaika and Minya rivers, the main divide line moves still further from the Baikal hollow. In the extreme, northern location the ridges reach 65–70 km at their bases, and their mean altitude above the sea level varies over the range 1650–1800 m above the sea level. This portion of the mountain massif now lies within 65 °N, which, together with an increase in continentality, influences vegetation. There occurs a decrease of the upper boundary of the mountain-taiga belt, accompanied by depletion of the vegetation communities of the goletzes. A shallow occurrence of acidic rocks of the Early Proterozoic is observed there. The landform is represented by a complex of alpinotype peaked ricges and high-mountain plateaus: the ancient peneplains occupying vast areas in the northwest of the territory. They are almost entirely covered with yernik and lichen (*Cladonia turgida, C. rangiferina*) -sphagnum (*Sphagnum girgensohnii*) mountain tundras in combination with great bilberries (*V. uliginosum*) -sedge (*Carex atrofusca, C. brunnescens, C. canenscens*) -sphagnum (*Sphagnum girgensohnii, Sh. Russowii, Sp. Fuscum*) communities of the Baikal-Dzhugdzhur formations: located in places of shallow occurrence of permafrost. The peaked ridges with steep slopes, due to a high strengthof rocks, look like monlithic rock plates devoid of vegetation. In the lower part, they are lined with colluvial large-block screes. Vegetation of screes is sparse and a little diverse according to its floristic composition: *Chamerion angustifolium*, *Bergenia crassifolia*, *Vaccinium myrtillus*, *Rhododendron aureum*. There singly occur sparse thickets of dwarf Siberian mountain pine with strong stems. Almost no alpinotype meadows were observed there.

4 Conclusion

By using the method of geobotanical mapping, it was possible to reveal the main distribution and development patterns of vegetation in the high mountains of Northwestern Cisbaikalia. It became clear that with the movement across the territory from south to north there occurs a decrease in indication capacity of vegetation for the composition of underlying material. The main criteria of its occurrence are represented by climate, geographical location and forms of mesorelief. A decrease in the height of the mountain ridge promotes the encroachment of forest species of flora onto the communities of the goletz belt. In the south of the territory, plants of the mountain tundras most often cushion-shaped, which indicates a rapid thawing of snow cover and exposure of the communities to strong winds. In the central and northern parts of the mountain ridge, large areas of rocky tracts are clad in snow for a long time, and vegetation cover is sparse there. Meadow communities are located below the nival boundary, and in the summer period they are provided with a long-lasting humidification by meltwater which is favorable for their habitat. Sparse vegetation of dwarf Siberian mountain pine in the south of the territory cover more humid or gentle slopes formed by acidic or basic material. In the central and northern parts of the mountain ridge, they are now represented by closed shrub thickets which refer to vegetation of the mountain-taiga belt.

Thus the findings reported above made it possible to model a complete picture of the modern status of high-mountain vegetation in Northwestern Cisbaikalia by illustrating its diversity and uniqueness. The results obtained in this research are relevant not only as the basis for a further monitoring of the modern status of vegetationand its diversity but also are basic to the study of geosystems of the Baikal region.

Acknowledgements. This work was done under a project of state assignment of V.B. Sochava Institute of Geography SB RAS no. AAAA-A17-117041910172-4.

References

1. Sochava, V.B., Timofeev, D.A.: Physical and geographical areas of Northern Asia. Reports of the Institute of Geography of Siberia and the Far East of the SB Academy of Sciences of the USSR **19**, 3–19 (1968). (in Russian)
2. Korolkova, E.E.: Medium-scale geobotanical mapping of the North-Western Baikal region taking into account the evolutionary and dynamic features of vegetation. Geobotanicheskoe kartografirovanie, 42–61 (2015). (in Russian)
3. Vladimirov, I.N.: Dynamics of the upper boundary of the forest on the Baikal ridge. Izvestiya Irkutskogo Gosudarstvennogo Universiteta. Series "Nauki o Zemle" 10, 46–56 (2014). (in Russian)

4. Belov, A.V., Lyamkin, V.F., Sokolova, L.P.: Cartographic study of biota. Oblmashinform, Irkutsk (2002). (in Russian)
5. Kononov, E.E.: Geological structure. Map. Ecological atlas of Lake Baikal basin. V.B. Sochava Institute of Geography SB RAS, Irkutsk (2015). (in Russian)
6. Kuzmin, V.A.: Soils of ecosystems of the northwest coast of Baikal. Geogr. Prir. Resour. **4**, 72–78 (1985). (in Russian)
7. Ladeishchikov, N.P.: Dynamics of the Baikal Basin. Nauka, Novosibirsk (1975). (in Russian)
8. Wyatt, B.K.: Vegetation mapping from ground, air and space – competitive or complimentary techniques? In: Vegetation Mapping (eds. Alexander, R., Millington, A.C.), pp. 3–15. J. Wiley & Sons, Ltd, Chichester (2000)
9. Millington, A.C., Alexander, R.: Vegetation mapping in the last three decades of the twentieth century. In: Alexander, R. Millington, A.C. (eds.) Vegetation Mapping: From Patch to Planet, pp. 321–331. John Wiley & Sons, Ltd, Chichester (2000)
10. Bezrukova, E.V., et al.: First high-resolution dated records of vegetation and climate changes on the Lake Baikal northern shore in the middle-late Holocene. Dokl. Akad. Nauk SSSR **411**(8), 1331–1335 (2006)
11. Bezrukova, E.V., et al.: Early hidh-resolution dated records of vegetation and climate change in the northern Baikal region in middle-late Holocene. SERIEs **411**, 254 (2006)
12. Franklin, J.: Mapping Vegetation from Landscape to Regional Scales. Vegetation Ecology, 486–508 (2013)
13. Sochava, V.B.: Vegetation Cover on Thematic Maps. Nauka, Novosibirsk (1979). (in Russian)
14. Belov, A.V., Sokolova, L.P., Fischer, E.E.: Cartographic modeling of vegetation of geosystems on an evolutionary-dynamic basis. Geograficheskie issledovaniya Sibiri **1**, 175–205. Geo, Novosibirsk (2007). (in Russian)
15. Anenkhonov, O.A., et al.: Determinant of plants of Buryatia. Republican Printing House, Ulan-Ude (2001). (in Russian)
16. Tyulina, L.N.: Dynamics of the Baikal Depression. Nauka, Novosibirsk (1975). (in Russian)
17. Makry, T.V.: Lichens of the Baikal Ridge. Nauka, Novosibirsk (1990). (in Russian)
18. Lukicheva, A.N.: Geobotanical studies and dynamics of shores and slopes on Lake Baikal. In: Tyulina, L.N., Ed. Nauka, Leningrad (1972). (in Russian)

Sustainable Development of Resource Regions: Challenges for Remote Areas of Siberia

Natalia Krasnoshtanova(✉)

V.B. Sochava Institute of Geography SB RAS, Ulan-Batorskaya, 1, 664033 Irkutsk, Russia
knesun@mail.ru

Abstract. The article presents the key issues and challenges relevant to remote areas of Siberia in the light of new economic development on the basis of a locally focused approach. The study is based on the results of field socio-geographical work carried out in February and March 2022 in three northern districts of Irkutsk Oblast, where a new oil and gas industry is actively developing. The findings are supplemented with data from the Federal State Statistics Service, documents of strategic planning at the regional and municipal levels, reports of heads of municipalities and other open sources concerning the issues of socio-economic development of the territories under study. The research results are systematized according to the main manifestations of the consequences of the new economic development: economy, society and ecology. The municipal settlements of the areas under study are grouped according to the degree and nature of the impact of the oil and gas industry on local socio-economic systems.

Keywords: resource regions · sustainable development · remote territories · new economic development

1 Introduction

The processes of new economic development, primarily related to the industrial extraction of natural resources, cover many different and new areas, moving further North into remote territories, having a significant impact on local socio-economic systems. Issues of sustainable development (SD) are becoming especially relevant for regions with a raw material economy based on non-renewable resources. Because the economy, based mainly on natural raw materials, with all the benefits received by various actors, is limited in its development by the volume of these raw materials and, as a rule, this fact causes significant damage to the natural environment. Moreover, if the main beneficiary is not always tied to a certain territory, then the social and environmental consequences of industrial development of raw materials have a direct impact on local actors. Hence, there is an imbalance between the interests of economic development, society and the preservation of the natural environment, which contradicts the main goals and principles of sustainable development. The urgency of developing an industrial policy for the development of the North and the Arctic of Russia is increasing, which will comprehensively

I. N. Vladimirov et al. (Eds.): RERSDNA 2022, SPEES, pp. 77–85, 2023.
https://doi.org/10.1007/978-3-031-28978-1_9

take into account the peculiarities of economic activity in this region and contribute to increasing the sustainability of socio-economic development [1, 2].

The oil and gas industry plays a special role in the economic development of remote areas of Eastern Siberia at the beginning of the 21st century. Large-scale industrial and infrastructure projects of federal significance are being implemented here. In the late 2000s, the construction of the Eastern Siberia – Pacific Ocean (ESPO) oil pipeline provided an opportunity to supply hydrocarbon raw materials to the market of the Asia–Pacific region (APR). At the same time, the presence of an oil-trunk pipeline contributed to the intensification of geological exploration and active industrial development of hydrocarbon deposits in Eastern Siberia, the exploitation of which was previously hindered by the inaccessibility of these territories and the lack of necessary infrastructure. In the 2010s, the construction of the Power of Siberia gas pipeline began, also focusing on the supply of raw materials to the Asia-Pacific market. Large infrastructure projects and the accompanying development of the oil and gas industry have a significant impact on the socio-economic development of the territories of their presence. On the one hand, there are new opportunities for the development of industrial territories, and on the other hand, they face new challenges and problems. Thus, the goal of the article is to identify the issues and challenges of socio-economic development of remote resource areas of the north of Irkutsk Oblast in the light of SD based on a locally oriented approach. To achieve the goal, it is necessary to solve the following tasks consistently. Firstly, we should review domestic and foreign scientific research, paying special attention to the latest scientific results in this field. Secondly, we try to collect heterogeneous socio-economic information in the form of qualitative and quantitative data on the studied territory, including through field socio-geographical work, and then to identify current problems and challenges facing the local communities by analyzing and synthesizing the collected materials.

The review of domestic and foreign scientific literature on the research topic reveals main directions that emphasize the relevance of the designated topic and determine the theoretical framework of the work:

-SD of the northern and Arctic regions [1–5];
-SD of the regions of Siberia and the Far East [6–8];
-SD of resource regions [2, 9];
-regional and local levels of SD [10–12];
-criteria and indicators of SD [8, 13, 14].

The field of SD of the northern and Arctic territories becomes particularly relevant in connection with the growing problems of climate change, which are most actively manifested in high latitudes. At the same time, the consequences of industrial development, which is actively moving to the north, have a negative impact on sensitive and vulnerable northern ecosystems. In terms of socio-economic development, special attention in the works of economists is paid to regions of resource specialization and different levels of SD are studied. An important direction is the study of the criteria and indicators of SD at the level of regions and settlements.

2 Materials and Methods

The data of the federal state statistics service, regional and municipal strategic planning documents, progress reports of the heads of municipalities, archival records and a number of other open sources relating to the socio-economic development of the areas under study served as basic materials for the article. Field socio-geographical studies were carried out in February and March 2022 in three northern districts of Irkutsk Oblast: Kirenskii, Katangskii and Ust-Kutskii, where the processes of industrial development of hydrocarbon resources are currently most actively taking place. 18 settlements of different status and distance from the places of industrial work were investigated. We used social science methods in data collection: semi-structured interviews, questionnaire and participant observation. 103 male and female respondents with an age range of 18 to 70 years old took part in the study, including the representatives of local administrations, social institutions (medical organizations, schools, kindergartens, libraries, cultural centers, etc.), industrial companies, business, as well as hunters, anglers, members of the indigenous communities of the North and other representatives of the local population. Comparative geographic, statistical and cartographic methods and general scientific methods of analysis and synthesis were used in data processing.

2.1 Study Area

The Ust-Kutskii district has the best transport and geographical position in comparison with other regions under study. Almost all types of transport are available here: railway (Baikal-Amur Mainline (BAM)); automobile (highways of federal importance "Vilyui" and regional "Ust-Kut – Uoyan"); air (air communication with the regional center and other cities, as well as helicopter services for industrial work); river (tanker transportation of goods along the Lena River, the port of Osetrovo); and pipeline (ESPO oil pipeline). The intersection of all these types of transport infrastructure is the city of Ust-Kut (regional center) as the base for the economic development of the north of Oblast and the neighboring Republic of Sakha (Yakutia). The main sectors of the region's economy - industry, transport, construction, forestry and trade are largely due to its favorable transport and geographical position.

The transport infrastructure of the neighboring Kirenskii district is less developed. There is virtually no railroad here. Only a small section of the BAM passes through the southern settlement of the municipality, but does not play a significant role in the socio-economic development of the entire district. There are no federal highways in this municipality. Regional roads are gravel. However, the main federal pipelines ESPO and the Power of Siberia pass through this area. In this regard, the structure of the modern economy of the region is determined by three main industries: transport and communications, oil and gas and timber industries. The areas of trade and agriculture are also important in the socio-economic development of the municipality.

The Katangskii district, located to the north of the Ust-Kutskii and Kirenskii districts, is the northernmost territory of the region. This is the only municipality in the territory of Irkutsk Oblast belonging to the regions of the Far North. It is the least-developed district in terms of transport. In addition, it belongs to the category of hard-to-reach areas, since there are no year-round roads, only winter roads. All-the-year-round communication with the settlements of the region is carried out by helicopters, and by an aircraft operating flights Irkutsk-Erbogachen with the regional center. The area is densely populated by Evenks, and there are about ten communities of indigenous peoples of the North engaged in traditional activities. At the same time, since the late 2010s, the oil and gas industry has been considered the main sector of the region's economy, but taking into account the employment of the local population, the organization of the public sector and hunting and fishing activities are the most significant. Table 1 presents a number of socio-economic indicators of the current development of the study areas.

Table 1. Indicators of socio-economic development of the areas under consideration (according to [15]).

Indicator	Ust-Kutskii	Kirenskii	Katangskii
The volume of shipped products of own production in 2021, billion RUB	339.0	71.7	277.4
Share of the mining sector in the total volume of shipped products in 2021, %	86.5	41.0	96.8
Population as of January 1, 2021, thou people	47.3	17.0	3.2
Population density, person/km^2	1.4	0.4	0.02
Municipal budget revenues per capita (including FS index) in 2020, thou rubles/person	36.5	29.5	34.5
Average salary in 2021, thou RUB	85.0	75.1	102.9

3 Results and Discussion

The issues and challenges of sustainable socio-economic development of the territories of the new industrial development of the north of Irkutsk Oblast, identified through a locally focused approach, in comparison with the results of other scientific studies in related areas, were systematized according to the main areas of manifestation: economy, society and ecology (see Table 2).

The problem of replenishing the budgets of settlements appears in the scientific literature from the very beginning of the transition of the Russian Federation to the path of sustainable development. It has always been noted, that in order to increase the capabilities of local administrative authorities to create a sustainable foundation for socio-economic development, it is necessary to decentralize the existing system of budgeting [27, 28]. At the same time, even in municipalities with a high level of their own incomes (in our case, settlements where direct hydrocarbon production is carried out), there are difficulties in managing the expenditure side of the budget caused by the limited powers of representatives of municipal governments [16].

The spatial and economic isolation of industrial development in remote areas of the North intensified during the COVID-19 pandemic, which coincided with the construction of the 2nd stage of the Power of Siberia gas pipeline in the Irkutsk Oblast:

"When the first pipe [ESPO oil pipeline] was being built, we were earning more, there were more orders, and we only had time to send cars. And now they have everything of their own, their own cars, they bring people, now a lot of things are passing by the city" (45 years old, local entrepreneur).

Table 2. Main issues and challenges of the territories of new industrial development.

Sphere	Form of manifestation	Interview examples
Economy	- Budgeting of municipalities (taxes and payments from a new sector of the economy) [9, 16]; - System of tender procurements and services (especially critical for remote territories) [17]; - Enclave process of new economic development of Northern territories [18, 19]; - Strengthening the monospecialization of the economy of settlements [17, 20]	*"Large enterprises work here, this is the oil and gas industry, it is flourishing and developing. They have their own transport logistics, their own roads"* (male, 65 years old, local resident)
Society	- COVID-19 pandemic (increased isolation, high incidence in mining areas) [16, 21]; - Shortage of personnel, including the leakage of qualified specialists to the oil and gas industry [22]; - Sociocultural interactions between the local population and industrial companies, including the field of preserving the traditional occupations of the indigenous peoples of the North [23]	*"Oil companies support us, but it didn't work out there in the pandemic year, but it's always like that. They also support participation in competitions, we go out and buy equipment, necessary materials"* (head of the Evenk center)
Ecology	- Anthropogenic transformation of natural landscapes [24]; - A growing network of technological roads (informal roads) [25]; - The consequences of climate change, which enhance the transformation processes in local ecosystems [26]	*"This is a disturbance factor for animals! Just imagine, they have lived here throughout ages, they had migration routes, and then bang - a drilling rig appeared, the river was filled up, profiles were cut, here a swamp appeared, there the swamp dried up, but everything is disturbed. Well, plus all these wildfires"* (male, 40 years old, local resident, hunter)

During the COVID-19 pandemic, industrial companies have been forced to minimize contact with third parties, including with the local population, to reduce the likelihood of outbreaks of infection of their workers. However, especially at the very beginning of the pandemic, it was in the areas of oil and gas production that outbreaks of the new coronavirus infection were noted, which created additional burdens on local medical institutions.

For the territories inhabited by the indigenous peoples of the North, in context of industrial development of natural resources, there are acute problems of preserving traditional forms of economic activity, which are closely interconnected with the state of the natural environment, which is experiencing negative consequences from industrial work. At the same time, oil and gas companies provide financial and material support to the local and indigenous population for the preservation of culture in the form of sponsorship directly or through socio-economic agreements.

The meaning of the above examples is very multifaceted (and even dual) and all of them are interconnected by emerging problems and challenges for local communities in the context of new industrial development. In the process of sustainable development, these problems and challenges become the most relevant, as complex non-linear relationships are revealed between economic development, society and the state of the natural environment of the northern territories. At the same time, the consequences of industrial development processes are highly differentiated in space. Based on the analysis and synthesis of the collected materials, the municipal settlements of the study areas were divided into four categories (see Fig. 1):

1. Territories with active hydrocarbon production. These are, as a rule, settlements with self-sufficient local budgets due to the payment of personal income tax by oil and gas companies. Interaction with industrial companies is carried out directly and through the district administration, in particular, educational programs are offered for the local population. However, there is a significant transformation of natural landscapes.
2. Territories of settlements with active geological exploration, as well as experimental and industrial exploitation of small deposits. A significant share of the revenues of the budgets of local settlements is formed at the expense of the oil and gas industry. This group also included hub settlements through which shift workers are delivered to their place of work. Here, the corresponding service sector is developing, but anyway accompanying social problems appear [29].
3. Territories for geological exploration and placement of service oil and gas pumping stations on main pipelines. The contribution to the formation of local budgets of settlements is insignificant, due to the small number of employees serving such facilities. The flow of shift workers is small, but at the same time, local residents perceive them mostly positively, since there is an additional opportunity to sell the products of subsidiary and peasant farms. As far as the infrastructural development in such territories, it becomes possible for local residents to use technological passages by passes of industrial companies, for example, along the ESPO oil pipeline. The negative impact on the natural environment is sectoral or has an areal extension (in the case of geological exploration), but it is temporary.
4. Territories of settlements where geological exploration and other works related to industrial development are absent or have been completed. However, this also includes transit areas of pipelines, but without pumping stations. Oil and gas companies interact with the local administration on the basis of socio-economic agreements. There is no income from the oil and gas industry in the budget of settlements.

These territories can receive socio-economic effects from the industrial sphere in the form of sponsorship through socio-economic agreements with local administrations and other indirect mechanisms [30]. Schematic diagram in Fig. 1 shows that quantitatively most of the settlements here belong to this 4th group. At the same time, the presence on their territory of federal infrastructure facilities (ESPO and the Power of Siberia gas pipeline under construction) does not have any significant socio-economic effect.

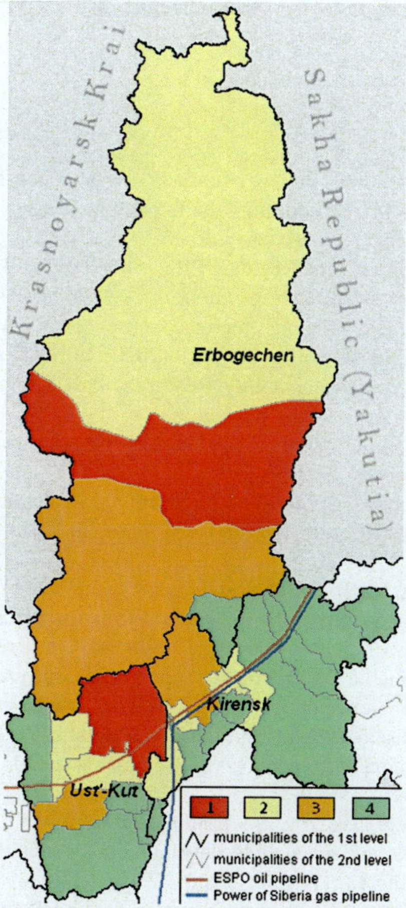

Fig. 1. Groups of municipal settlements in the north of Irkutsk region by the nature of the manifestation of the processes of new industrial development (explanations in the text).

4 Conclusion

In the context of sustainable development, new challenges and issues faced by municipalities in the north of Irkutsk Oblast become especially relevant, as all areas of local socio-economic systems (economy, society and natural environment) are affected. The locally focused approach enabled the identification of the barriers to sustainable socio-economic development at the level of settlements, which in fact become constraints on the transition to sustainability at the regional and national levels. The results obtained will form the basis for the proprietary technology of sustainable development indicators at the local level for the northern territories, with the aim of subsequent assessment of the sustainable development potentials of the territories under consideration.

Acknowledgments. The study was supported by a grant from the Russian Science Foundation No. 21-78-00057 "The Northern regions of new development: prospects for sustainable development".

References

1. Pilyasov, A.N.: Arctic industrial policy: not assets and branches, but resources and corporations. The North and the Market: Forming the Economic Order **1**, 41–58 (2020). https://doi.org/10.25702/KSC.2220-802X.5.2018.61.180-189
2. Uss, A., et al.: How to Increase the Regional Effects of Resource Projects. ECO. All-Russian Economic Journal **2**, 27–46 (2022). https://doi.org/10.30680/ECO0131-7652-2022-2-27-46
3. Nilsson, A.E., Larsen, J.N.: Making regional sense of global sustainable development indicators for the arctic. Sustainability **12**, 1027 (2020). https://doi.org/10.3390/su12031027
4. Petrov, A.N., et al.: Arctic Sustainability Research: Past Present and Future. Routledge, London (2017)
5. Graybill, J.K., Petrov, A.N.: Arctic Sustainability Key Methodologies and Knowledge Domains. Routledge, London (2020)
6. Baklanov, P.Y.: Sustainable development of the Primorsky Regions: geographical and geopolitical factors and limitations. Baltic Region **1**(14), 4–16 (2022). https://doi.org/10.5922/2079-8555-2022-1-1
7. Mutovin, S.I., Sud'in, K.N.: Tools for Sustainable Development of the Northern Territories: The Practice of Regional Studies. Siberian Federal University, Krasnoyarsk (2014)
8. Abalakov, A.D., Bazarova, N.B., Novikova, L.S.: Siberia, Irkutsk region: zoning of sustainable development. In: Sustainable Development in East Asia: Current Ecological, Geographical and Socio-Economic Problems, pp. 12–15. IG SB RAS, Irkutsk (2018)
9. Sevastyanova, A.E., Yatsenko, V.A.: Barriers of sustainable development of municipalities with resource specialization of the economy. J. New Econ. **4**, 174–191 (2020). https://doi.org/10.29141/2658-5081-2020-21-4-9
10. Vlasova, T., Petrov, A.N., Volkov, S.: Rethinking sustainability monitoring in the arctic by linking resilience and sustainable development in socially-oriented observations: a perspective. Sustainability **13**, 177 (2021). https://doi.org/10.3390/su13010177
11. Oosterhof, P.D.: Localizing the sustainable development goals to accelerate implementation of the 2030 agenda for sustainable development. The Governance Brief **33**, 1–14 (2018)
12. Tishkina, T.M.: Problems of transition to sustainable development at the municipal level. Economy: Yesterday, Today, Tomorrow **8**(8), 132–141 (2018)
13. Zemtsov, S., et al.: Ecological efficiency and sustainable regional development in Russia during the 20 years of resource-based growth. Economic policy **2**(15), 18–47 (2020). https://doi.org/10.18288/1994-5124-2020-2-18-47
14. Bobylev, S.N.: New economic models and indicators of sustainable development. The Economic Revival of Russia **3**(61), 23–29 (2019)

15. Federal State Statistics Service. https://www.gks.ru/scripts/db_inet2/passport/table. Last accessed 2022/08/20
16. Krasnoshtanova, N.E.: Features of administrative management in the areas of new industrial development of the North: problems and advantages of socio-economic development. The North and the Market: the Formation of an Economic Order **2**, 82–96 (2022). https://doi.org/10.37614/2220-802X.2.2022.76.007
17. Saxinger, G., et al.: Boom back or blow back? Growth strategies in mono-industrial resource towns–'east'and 'west'. In: Settlements at the Edge: Remote Human Settlements in Developed Nations, pp. 49–74. Edward Elgar Publishing, Cheltenham (2016)
18. Shmat, V.V.: "Center" turned out to be right, because he took more rights? ECO. All-Russian Economic Journal **7**, 60–77 (2013)
19. Pilyasov, A.N., Zamyatina, N.Y.: Development of the North 2.0: challenges of making a new theory. Arctic and North **34**, 57–76 (2019). https://doi.org/10.17238/issn2221-2698.2019.34.57
20. Frynas, J.G.: The false developmental promise of corporate social responsibility: evidence from multinational oil companies. Int. Aff. **3**(81), 581–598 (2005). https://doi.org/10.1111/j.1468-2346.2005.00470.x
21. Panin, A.N., Rilskiy, I.A., Tikunov, V.S.: Spatial patterns of COVID-19 distribution in Russia and the world: cartographic analysis. Bulletin of the Moscow University. Series 5. Geography **1**, 62–77 (2021)
22. Kuklina, V.V., Krasnoshtanova, N.E.: The urbanization of indigenous peoples of northeastern Siberia. In: Laruelle, M. (ed.) New Mobilities and Social changes in Russia's Arctic Region, pp. 104–129. Routledge, New-York (2017)
23. Tulaeva, S., Tysiachniouk, M.: Between oil and reindeer: benefit sharing agreements between oil companies and indigenous people in Russian Arctic and Subarctic Regions. J. Econ. Sociol. **18**(3), 70–96 (2017). https://doi.org/10.17323/1726-3247-2017-3-70-96
24. Opekunova, M.G., et al.: Assessment of the transformation of the natural environment in the areas of hydrocarbon raw materials development in the north of Western Siberia. Siberian Ecol. J. **1**, 122–138 (2018). https://doi.org/10.15372/SEJ20180111
25. Kuklina, V., et al.: Mobilizing benefit-sharing through transportation infrastructure: informal roads, extractive industries and benefit-sharing in the Irkutsk oil and gas region Russia. Resources **9**(3), 21 (2020). https://doi.org/10.3390/resources9030021
26. Report on the peculiarities of the climate in the territory of the Russian Federation for 2016. Roshydromet, Moscow (2017)
27. Petrova, T.A.: Some aspects of the formation of sustainable development of municipalities. Innovation **3**, 91–93 (2008)
28. Tokaev, N.H.: Formation and self-sufficiency of income support of municipal budgets. Terra Economicus **3**(10), 118–122 (2012)
29. Taylor, A.J., Carson, D.B.: It's raining men in Darwin: gendered effects from the construction of major oil and gas projects. Journal of Rural and Community Development **1**(9), 24–40 (2014)
30. Tysiachniouk, M., Petrov, A.: Benefit sharing in the arctic energy sector: perspectives on corporate policies and practices in Northern Russia and Alaska. Energy Res. Soc. Sci. **39**, 29–34 (2018). https://doi.org/10.1016/j.erss.2017.10.014

Problems of Population Reproduction in the Cities of Siberia

Ya. A. Leshchenko(✉) and A. A. Lisovtsov

East-Siberian Institute of Medical and Ecological Research, Bldg. 3, Microdistrict 12a, P.O. Box 1170, Angarsk 665827, Russia
yaleshenko@gmail.com

Abstract. The main problems of natural and mechanical reproduction of the population of the cities of the Siberian Federal District in the first twenty years of the 21st century are characterized. Insufficient (narrowed) natural reproduction of the population was revealed in large administrative centers - Novosibirsk, Krasnoyarsk and Irkutsk. There was a positive balance of migration during almost the entire observation period in Krasnoyarsk. In Irkutsk, a negative balance of migration was recorded between 2000 and 2007 and 2016 and 2020. In Novosibirsk, a negative balance of migration was registered between 2002 and 2005. The features of the demographic development of the industrial cities of Siberia are demonstrated on the example of the city of Angarsk. During the first twenty years of the 21st century, the processes of natural and mechanical reproduction of the population in large administrative and industrial cities of Siberia are unfavorable. Indicators of mechanical reproduction of the population are characterized either by the predominance of migration outflow (Irkutsk), or by a consistent decrease in migration inflow, started in 2015 (Krasnoyarsk and Novosibirsk). Ensuring an increase in the demographic potential of Siberian cities by increasing natural reproduction is not possible due to the small number of contingents of the population of reproductive age. The only possible way to solve this problem is migration policy. It is necessary to move from the compensatory model of the economy to the model of advanced development of human potential.

Keywords: cities of Siberia · demographic potential · natural and mechanical reproduction · sustainable development

1 Introduction

In recent decades, the issues of the demographic potential formation are of particular importance among the issues of sustainable development of regions in many countries [1–3]. This problem is most acute in Siberia and the Far East as vast sparsely populated territories of great economic, social and geopolitical importance.

The sustainable development of regions is possible only with the creation and improvement of various types of labor-intensive and science-intensive production, which can only be provided by a sufficient number and quality of the population, united by a complex system of economic, social, logistical and industrial relations [4, 5]. The

I. N. Vladimirov et al. (Eds.): RERSDNA 2022, SPEES, pp. 86–94, 2023.
https://doi.org/10.1007/978-3-031-28978-1_10

demographic potential concentrated in the network of modern cities is required for the successful development of the eastern macroregions of Russia, including the Siberian Federal District (SFD), from the standpoint of the modern technological order. The cities and the administrative centers of the regions, as well as large industrial cities created after the Second World War are considered as places of concentration of fuel, energy, and timber industry complexes, machine-building, chemical, metallurgical and other enterprises. Such cities have become an important link in solving economic, political and defense problems. At the same time, the task of creating a supporting frame for settlement in the east of the country was also solved, which is of great geopolitical importance.

With the onset of the socio-economic and political crisis in the late 1980s and early 1990s, the problems of the country's urban and demographic development became much more acute. The number of urban (as well as general) population of Russia began to decline steadily. The change in the dynamics of the urban population is one of the signs of trouble in the development of Russian cities, especially those located in Siberia.

In the context of the growing struggle for resources in the world, depopulation caused by natural loss and migration outflow of the population from the territory of Siberia has a negative socio-economic and geopolitical significance. The decline in the number of Russians and their replacement by migrants carries significant ethnocultural, socioeconomic, and geopolitical risks [6–8].

The goal of the work is to characterize the main problems of population reproduction in the cities of the Siberian Federal District (SFD) in the first twenty years of the 21st century.

2 Materials and Methods

The objects of study are the administrative centers of the regions and young industrial cities of Siberia. Demographic processes in administrative centers are considered on the example of the cities of Novosibirsk, Krasnoyarsk and Irkutsk. At the beginning of 2022 the population in Novosibirsk was 1621330 people, in Krasnoyarsk – 1103023 people and in Irkutsk – 617249 people. The features of the demographic development of young Siberian cities are demonstrated on the example of Angarsk, a large industrial center of Irkutsk Oblast.

The natural movement (reproduction) of the population is estimated by the nature and ratio of birth and death rates. The mechanical movement of the population is estimated according to the characteristics of intraregional, interregional and external migration (concerning the countries of the Commonwealth of Independent States (CIS) and other foreign countries). The coefficients of migration and the balance of migration are considered as the final criteria for migration processes. Statistical materials are obtained from the databases of the Federal State Statistics Service of the Russian Federation (Rosstat) and the Center for Demographic Research, as well as the territorial bodies of Rosstat for the Krasnoyarsk Territory, Novosibirsk and Irkutsk Regions.

3 Results

3.1 Reproduction of Population in the Administrative Centers of Siberian Regions

Natural Reproduction. The dynamics of vital statistics (VS) of large Siberian cities – Novosibirsk, Krasnoyarsk and Irkutsk in the period from 2002 to 2021 is presented in Table 1.

Table 1. Indicators of the natural and mechanical movement of the population of Novosibirsk, Krasnoyarsk and Irkutsk in the period from 2000 to 2021 (per 1000 population).

Year	Irkutsk				Krasnoyarsk				Novosibirsk			
	CBR	CDR	INMP	NMR	CBR	CDR	INMP	NMR	CBR	CDR	INMP	NMR
2000	-	-	-	−5.6	-	-	-	52.8	-	-	-	17.7
2001	-	-	-	−27.4	-	-	-	35.1	-	-	-	14.4
2002	11.5	15.5	−4.0	−39.7	10.7	14.0	−3.3	49.5	9.3	14.6	−5.3	−12.9
2003	12.2	15.8	−3.6	−47.2	10.9	13.7	−2.8	31.5	9.7	14.5	−4.8	−30.5
2004	12.5	15.3	−2.8	−33.3	10.9	12.8	−1.9	62.8	10.2	14.5	−4.3	−16.7
2005	12.0	15.2	−3.2	−45.5	10.9	13.3	−2.4	61.5	10.1	15.1	−5.0	−15.5
2006	12.2	13.8	−1.6	−18.1	11.1	11.9	−0.8	76.4	9.9	14.0	−4.1	1.9
2007	13.5	12.6	−0.9	−9.7	11.4	11.7	−0.3	104.6	10.5	13.9	−3.4	23.4
2008	14.4	12.7	1.7	42.8	12.7	11.4	1.3	107.7	11.7	13.4	−1.7	64.5
2009	14.8	12.9	1.9	5.4	13.4	10.8	2.6	127.3	12.4	12.9	−0.5	89.3
2010	14.2	12.5	1.7	53.9	13.5	11.0	2.5	139.3	12.9	12.8	0.0	68.8
2011	14.9	11.9	3.0	102.4	13.2	10.8	2.4	162.6	12.5	12.2	0.3	156.1
2012	15.2	11.9	3.3	103.6	14.1	10.8	3.3	154.7	13.3	12.2	1.1	152.9
2013	15.3	11.7	3.6	76.4	14.2	10.4	3.8	147.1	13.5	12.0	1.5	140.2
2014	14.8	11.9	2.9	85.9	14.7	10.3	4.4	115.8	13.6	11.8	1.8	104.5
2015	16.3	11.8	4.5	7.7	15.4	10.3	5.1	87.7	14.6	11.6	3.0	78.2
2016	16.0	11.2	4.8	−43.0	15.1	10.2	4.9	99.5	14.1	11.7	2.4	93.6
2017	14.6	10.8	3.8	−35.5	12.8	10.1	2.7	44.7	12.5	11.5	1.0	52.4
2018	13.9	10.8	3.1	−37.1	12.2	10.4	1.8	22.8	12.0	11.8	0.2	30.0
2019	12.0	10.8	1.2	−11.2	11.0	10.2	0.8	−22.1	10.9	11.4	−0.5	51.6
2020	12.4	13.5	−1.1	−40.2	10.6	12.6	−2.0	13.9	10.5	14.3	−3.8	6.2
2021	12.1	15.8	−3.7	-	10.2	14.4	−4.2	-	10.4	15.5	−5.1	-

Notes: CBR – crude birth rate; CDR – crude death rate; INMP – indicator of the natural movement of the population; NMR – net migration rate.

Over the entire observation period, the values of the total fertility rate were in Novosibirsk 9.3–14.6 %, in Krasnoyarsk – 10.7–15.4 % and in Irkutsk – 11.5–16.3 %, which, according to the criteria of the World health care organization corresponds to a low birth rate. The highest values of the overall mortality rate were noted: in Novosibirsk in 2002–2009 and in 2020–2021 (12.9–15.5 %); Krasnoyarsk – in 2002–2005 and in 2020–2021 (12.6–14.4 %); Irkutsk – in 2002–2007 and in 2020–2021 (13.5–15.8 %). Natural population decline was registered: in Novosibirsk in 2002–2009 and in 2019–2021, in Krasnoyarsk and Irkutsk - in 2002–2007 and in 2020–2021.

When considering all three indicators simultaneously, we can conclude that the character of the natural reproduction of the population in the large Siberian cities in the first two decades of the 21st century is insufficient (narrowed), even with the positive changes in 2008–2019.

Mechanical Reproduction. The structure of the mechanical movement of the population changed significantly depending on the socio-economic situation in the country and individual regions.

The combined interaction of three types of migration processes led to the following picture in the formation of the total balance of migration. The most favorable indicators of migration activity were noted in Krasnoyarsk, where there was a positive balance of migration (per 10,000 population) over the entire observation period (13.9–162.60/000) (Table 1). The processes of intra-regional migration made the main contribution to the overall positive balance in 2000–2016. The least favorable situation was in Irkutsk, where the negative balance of migration was registered in 2000–2007 (−5.6 – −47.20/000) and in 2016–2020 (−11.2 – −43.00/000). The processes of intra- and inter-regional migration, which changed in opposite directions, had the greatest influence on the situation in Irkutsk: the intra-regional migration increased largely the balance of migration, while the inter-regional migration decreased it.

Since 2007, the influx of migrants from the CIS and Baltic countries has become increasingly important in Krasnoyarsk and Irkutsk. Since 2017, this area has become dominant in the formation of the overall balance in Krasnoyarsk.

Novosibirsk occupied an intermediate position: a negative migration balance was registered in 2002–2005 (−12.9 – −30.50/000). The processes of interregional migration, as well as the processes of immigration from the CIS countries, which have taken leading positions since 2015, had the greatest significance for the formation of the migration balance in Novosibirsk.

3.2 Population Reproduction in Young Industrial Cities of Siberia

There are 158 cities on the territory of modern Russia suitable, under the definition of 'young city' with certain reservations. Most of them – 36 cities (or 23%) – are concentrated within the Siberian Federal District (SFD) [9].

We demonstrate the features of the demographic development of young cities in Siberia using the example of Angarsk, a large industrial center of Irkutsk Oblast. Angarsk is a typical example of cities built to develop the military-industrial potential in the eastern regions of the country.

The population of Angarsk was formed, especially in the initial, most rapid period of growth, due to the migration influx, which was generated by a meaningful spatial strategy of the state for the development of the eastern regions of the country and the development of military-industrial potential in them. Since the foundation of the city (1951), the population has been constantly increasing, reaching a maximum in 1991 (269,181 people). In the period after the collapse of the USSR and the deployment of a systemic socio-economic and civilizational crisis, starting from 1992, the natural population growth was replaced by its decline. During all subsequent years, the city's population gradually declined (slowly at first then very rapidly). The main prerequisites for these adverse events were the changes that occurred in the processes of natural movement of the population: a pronounced increase in overall mortality and a rapid decrease in the birth rate.

The process of population aging began to develop rapidly. If the demographic age of the population of Angarsk corresponded to the stage of 'demographic youth' for 30 years (1950–1980), then in 1989 and 1990 it was characterized as 'old-age threshold', in 1991–1997 – as 'initial level of demographic ageing', in 1998–2000 – as 'medium level of demographic ageing', and since 2001 – as 'high level of demographic ageing' [10] (Table 2). The aging process causes an increase in the death rate, increases the burden on the working population, as well as on health and social protection systems, and complicates the provision of the economy with labor resources.

Table 2. Changes in the number and demographic age of the population of Angarsk.

Year	Number of population (people)	Proportion of people over 60 (%)	Demographic age stage
1959	134390	3.8	Demographic youth
1970	203310	5.9	Demographic youth
1979	240410	7.6	Demographic youth
1989	265615	11.0	Old-age threshold
1990	268059	11.7	Old-age threshold
1991	269181	12.1	Initial level of demographic ageing
1997	267023	13.7	Initial level of demographic ageing
1998	266502	14.5	Medium level of demographic ageing
2000	264234	15.9	Medium level of demographic ageing
2001	263347	16.3	High level of demographic ageing
2005	246815	16.6	High level of demographic ageing

(continued)

Table 2. (*continued*)

Year	Number of population (people)	Proportion of people over 60 (%)	Demographic age stage
2009	241483	17.4	High level of demographic ageing
2019	225060	21.6	High level of demographic ageing
2021	222855	22.1	High level of demographic ageing

In addition to demographic age, to characterize the composition of the population, the age structure is important, determined by the ratio of three age groups of the population: children (0–14 years), parents (15–49 years) and grandparents (50 years and older). The age structure of the population between 1959 and 1979 should be characterized as progressive. The share of children, parents and grandparents in these years was 22.7–26.7%, 60.8–64.7% and 8.6–16.5%, respectively. In 1989–1995, the type of age structure approached the stationary one (17.5–22.7%, 56.6–55.1%, 22.6–25.9%, respectively). In 1996–2001, it was characterized as approaching regressive, and in 2002–2009 and in subsequent years as regressive (14.8–18.2%, 47.8–56.6%, 25.9–34.0%, respectively).

4 Conclusion

Siberia is still a vast sparsely populated territory with many natural resources, the value of which is rapidly increasing at the present stage of world development (forest, energy resources, fresh water, land suitable for agricultural use, etc.). It is obvious that the demographic potential of Siberia is insufficient to create a developed economic and settlement structure, to develop the natural resources available here.

The materials of the study indicate the ongoing natural decline in the population of Siberian cities (the situation has not significantly changed by the very low rates of natural increase were temporarily observed between 2007 and 2018). The indicators of mechanical reproduction of the population also deserve a pessimistic assessment, since they are characterized either by the predominance of the migration outflow of the population (Irkutsk), or by a consistent decrease in the migration inflow of the population, starting from 2015 (Krasnoyarsk and Novosibirsk). Kashnitsky [11] reports that the capitals of the regions of Siberia have a relative migration attractiveness for young people, due to the possibility of obtaining a high-quality university education [12, 13], and also generally serve as places of attraction for interregional migrants, especially those born in border areas of the Siberian Federal District [14]. Moreover, the farther the settlement is from the regional capitals, regardless of its status (village, urban-type settlement or city) and size, the greater the outflow of young people from it [15]. The increase in the population of regional centers that we noted due to intraregional migration is typical for the whole country as a whole [16, 17]. The increase in the structure of mechanical reproduction of the influx of migrants from the CIS countries cannot be considered a positive trend, since these contingents are people of a low educational level and low qualifications. Other researchers [11, 18] share the same point of view. In addition, immigration external to cities can be accompanied by an increase in sociocultural risks [19].

Thus, during the first two decades of the 20th century, the processes of natural and mechanical reproduction of the population in large administrative and industrial cities of Siberia are unsatisfactory, if considered from the standpoint of ensuring the sustainable development of regions based on the modern technological order. Moreover, according to Zemtsov et al., many indicators of socio-economic development returned to the values of the mid-2000s after the crisis of 2015–2016. In 2020, the values of most indicators largely decreased than GRP per capita. In addition, a significant gap remains between the leading and lagging regions [5].

The most difficult task is to improve the situation in the natural reproduction of the population. Measures to stimulate the birth rate, carried out within the framework of the national project “Demography”, seem to be insufficient. A radical improvement in the entire social sphere is required. In any case, it will not be possible to ensure the necessary increase in the demographic potential of Siberian cities in a relatively short time by increasing the level of natural reproduction. The main reason is the insignificant number of the population of reproductive age, which, moreover, have decreased in the post-Soviet period.

Migration policy remains the only possible way to solve this problem. According to the Concept of the State Migration Policy of the Russian Federation for 2019–2025, the goal is “to create a migration situation that contributes to solving problems in the socio-economic, spatial and demographic development of the country, as well as in protection and preservation of the Russian culture” [20]. However, it should be recognized that the implementation of the above tasks is carried out unsatisfactorily.

There is an obvious need to develop an effective strategy for the sustainable development of the Siberian macroregion. The existing programs for the development of the regions of the Siberian Federal District actually solve the problem of creating the so-called compensatory model of the economy, which is aimed at reducing the severity of the main economic problems. The development of Siberia within the framework of these programs is one-sided (unsustainable and unbalanced development): the state is concerned exclusively with the growth of the economy, and people are considered only as its addition. Funds invested from the budget for solving the housing problem, social security, the development of health care and education systems, and the leisure of the population, do not compensate for the difficulties of life in these territories. The state of affairs is similarly characterized in other research works on the study of social problems in Siberia and the Far East [14, 16, 21]. Indigenous people leave Siberia, striving to get the same opportunities as in more prosperous regions of the country. In the territories under consideration, local natives form the core of the settlement of the population [14]. According to Shvorina et al., the proportion of those born in the Krasnoyarsk Krai, Irkutsk and Novosibirsk Oblasts, but living in other regions, was 19.8–23.5% in 2010 [14].

At the same time, because of the “western drift” of the population, ten constituent entities of the Russian Federation, located mainly in the European part of Russia

(Moscow and Moscow Oblast, St. Petersburg and Leningrad Oblast, Khanty-Mansiysk and Yamalo-Nenets Autonomous Okrugs) concentrated 87% of all entering interregional and 65% of foreign labor migrants in 2012–2013. Eighty percent of migrants changed their permanent place of residence and moved to the cities of these subjects [16].

The situation will not change significantly until a fundamentally different economy, cultural and recreational environment, and transport schemes oriented to the diverse needs of the population are formed in Siberia. There is an urgent need to move from a compensatory model of the economy to a model of rapid development of human potential, in which the creation of a socially and demographically oriented economy should come to the fore. Despite the existing difficulties, it is necessary to create incentives to attract and retain migrants of the appropriate composition (by age, gender, qualifications and cultural orientation) in Siberian cities, as was done in the Soviet period, when people were attracted to Siberia with high salaries, low taxes, and significant funds were invested in infrastructure development.

Only the solution of the above tasks will ensure the effective strategic development of the Siberian macroregion, where it will be possible to develop various types of innovative and science-intensive production in order to meet the criteria of the current level of civilizational development.

References

1. Kulczyk-Dynowska, A., Stacherzak, A.: The impact of a city on its environment: the prism of demography and selected environmental and technical aspects based on the case of major lower Silesian cities. Sustainability **14**(11), 6393 (2022). https://doi.org/10.3390/su14116393
2. Takashi, A.: Confronting future urban perforation: spatial analysis of districts in Japan with potential for being sparsely inhabited. Cities **122**, 103515 (2022). https://doi.org/10.1016/j.cities.2021.103515
3. Rodríguez-Domenech, M.Á.: Medium-sized cities facing the demographic challenge in Spain's low-density regions through citizen participation projects. Int. J. Environ. Res. Public Health **19**(9), 5303 (2022). https://doi.org/10.3390/ijerph19095303
4. Schackmar, J., Fleschurz, R., Pallagst, K.: The role of substitute industries for revitalizing shrinking cities. Sustainability **13**(16), 9250 (2021). https://doi.org/10.3390/su13169250
5. Zemtsov, S.P., Komarov, V.M., Barinova, V.A.: Uncovering new economy potential of Russian regions on the basis of the last 20 years dynamics' analysis. Ekonomika regiona [Economy of regions] **18**(1), 92–104 (2022). https://doi.org/10.17059/ekon.reg.2022-1-7
6. Krasnov, A.: Nezakonnaya migratsiya kak ugroza gosudarstvennosti [Illegal migration as a threat to statehood]. Bezopasnost' Yevrazii [Safety of Eurasia] **2**, 208–214 (2010). (in Russian)
7. Soboleva, S.V., Smirnova, N.E., Chudayeva, O.V.: Population migration in the border regions of Siberia. ECO **8**, 18–34 (2014). (in Russian)
8. Soboleva, S.V., Grigoriev, Y., Smirnova, N.E., Chudayeva, O.V.: Features of population formation of Siberian border territories. ECO **11**, 20–36 (2014). (in Russian)
9. Bol'shakova, A.A.: Molodyye goroda Irkutskoy oblasti kak potentsial osvoyeniya Vostochno-Sibirskogo regiona i resurs realizatsii strategii prostranstvennogo razvitiya Rossii [Young cities of the Irkutsk region as the potential for the development of the East Siberian region and the resource for the implementation of the spatial development strategy of Russia]. In: Grabel'nykh, T.I., Reshetnikov, V.A. (science eds.) Moderniztsia i konstruirovanie strategiy regionalnogo i gorodskogo razvitia [Modernization and construction of strategies of regional

and urban development]: collection of scientific papers, pp. 151–165. ISU, Irkutsk (2011). (in Russian)
10. Leshchenko, Y.A.: Problems of development of the population of a Siberian town. RIO NTs RVKh VSNTs SO RAMN, Irkutsk (2010). (in Russian)
11. Kashnitsky, I., Mkrtchyan, N., Leshukov, O.: Interregional Migration of Youths in Russia: A Comprehensive Analysis of Demographic Statistics. Educational Studies. Moscow **3**, 169–203 (2016). https://doi.org/10.17323/1814-9545-2016-3-169-203
12. Agasisti, T., Bianco Dal, A.: Determinants of college student migration in Italy: empirical evidence from a gravity approach. Soc. Sci. Res. Netw. (2007). https://doi.org/10.2139/ssrn.1063481
13. Ciriaci, D.: Does university quality influence the interregional mobility of students and graduates? The Case of Italy. Reg. Stud. **48**(10), 1592–1608 (2014). https://doi.org/10.1080/00343404.2013.821569
14. Shvorina, K.V., Faleychik, L.M.: Main directions of migration mobility in the Siberian and Far Eastern federal districts. Ekonomika regiona [Economy of regions] **14**(2), 485–501 (2018). https://doi.org/10.17059/2018-2-12. (in Russian)
15. Mkrtchyan, N.V.: Migration of youth in regional centers of Russia at the end of 20th - early 21st century. Izvestiya Rossiiskoi Akademii Nauk. Seriya geograficheskaya **6**, 19–32 (2013). (in Russian)
16. Nefedova, T.G.: Migration mobility of population and otkhodnichestvo in modern Russia. Izvestiya Rossiiskoi Akademii Nauk. Seriya geograficheskaya **3**, 41–56 (2015). (in Russian)
17. Karachurina, L.B., Mkrtchan, N.V.: Role of migration in enhancing contrasts of settlement pattern at municipal level in Russia. Izvestiya Rossiiskoi Akademii Nauk. Seriya geograficheskaya **5**, 46–59 (2016). https://doi.org/10.15356/0373-2444-2016-5-46-59. (in Russian)
18. Akumova, N.V., Ivanov, G.S.: Current status of labor migration in the context of the overall migration in Russia. Law and Education **10**, 114–122 (2013). (in Russian)
19. Gafiatulina, N., Samygin, S.I., Rachipa, A.V.: Social health of Russian youth in the context of regional migration processes: threats to national security. Humanities of the south of Russia **19**(3), 59–71 (2016). (in Russian)
20. Kontseptsiya gosudarstvennoy migratsionnoy politiki Rossiyskoy Federatsii na 2019–2025 gody [The concept of the state migration policy of the Russian Federation for 2019–2025]. http://www.consultant.ru/cons/cgi/online.cgi?req=doc&base=LAW&n=310139&fld=134&dst=100018,0&rnd=0.7377705270328321#07478829695252702. (last accessed 2019/03/10). (in Russian)
21. Motrich, Y.L., Nayden, S.N.: Vliyaniye sotsial'no-demograficheskoy situatsii na trudovuyu migratsiyu: dal'nevostochnyy vektor [The influence of the socio-demographic situation on labor migration: Far Eastern Vector]. Problemy prognozirovaniya **116**(5), 134–146 (2009). (in Russian)

Evaluation of the Transportation in Investment Environment from Land-Sea-Air Coordination: A Case Study of Northeast China

Li Fujia[1,2](✉), Guo Kexin[1,2], and Cheng Hao[1]

[1] Institute of Geographic Sciences and Natural Resources Research, Chinese Academy of Sciences, 11A, Datun Road, Chaoyang, Beijing 100101, China
lifj@igsnrr.ac.cn

[2] College of Resource and Environment, University of Chinese Academy of Sciences, Beijing, China

Abstract. Transportation network is an important link for economic cooperation between cities. It is of great significance to build a coordinated transportation network of land, sea and air to improve the service level of the international logistics supply chain and promote the coordinated development of land, sea and air economy. This paper takes northeast China as the research area, from the perspective of Land-Sea-Air Coordination. We consider comprehensive traffic accessibility in the evaluation of transportation in the investment environment. Qualitative indicators relating to convenience, construction scale and service level were selected to build a transportation evaluation index system in the investment environment using the entropy model. And then, we measure transportation in the investment environment and regional differences in northeast China. The results can deepen the scientific understanding of transportation in the investment environment in northeast China. The sea, land, and air resources can flow smoothly and complementary. Based on it, we can realize high-quality and efficient transportation in the investment environment construction. This study is of great significance to reveal the role of comprehensive transportation accessibility in the investment environment of transportation and improve the evaluation system of the investment environment.

Keywords: Investment Environment · Land-sea-air Coordination · Northeast China

1 Introduction

As economic globalization progresses, the competition between regions to attract capital becomes more and more intense. Investment environment is an important reflection of a country or region's economic competitiveness, and a good investment environment is a powerful magnetic field for a country or region to attract investment [1]. Research on investment environment evaluation began in the 1960s [2], the main methods are

I. N. Vladimirov et al. (Eds.): RERSDNA 2022, SPEES, pp. 95–104, 2023.
https://doi.org/10.1007/978-3-031-28978-1_11

investment hot comparative analysis, investment environment scoring method, static and dynamic gravity model, factors analysis method, etc. [3–9].

Faced with the current problems of land resource scarcity, population expansion and environmental degradation, China has formed a complex system in which the sea, land and air systems interact and coordinate with each other to promote sustainable economic development [10–13]. It takes advantage of the innate resource advantages and vast development space of the ocean and air systems, and promotes the land economy sectors extended to the ocean and air. To better meet the global resource restructuring and enhance the efficiency of investment. We construct a comprehensive evaluation index system of the investment environment from four aspects: resource utilization, economic industry, social development and investment convenience, based on the perspective of Land-sea-air coordination. It can improve the investment environment evaluation, and deepen the understanding of the transportation environment for regional and national investment and provide a quantitative evaluation method and conclusion basis for regional and national investment policy and decision making.

2 Objects and Methods

The three northeastern provinces of China include Heilongjiang, Jilin, and Liaoning provinces. The northeast region has always faced problems such as backward infrastructure, low level of industrial technology and serious environmental pollution. In recent years, the three northeastern provinces have actively responded to the call of the national land and sea integration strategy [14, 15]. It has played a great role in supporting the promotion of local development. So, we aim to quantitatively evaluate the current investment environment in Northeast China under land-sea-air coordination. It can provide theoretical reference and technical support for the planning and future development issues in Northeast China.

From the perspective of land–sea–air coordination, 33 quantitative indicators are selected, considering the correspondence and data integrity of indicators at each level. GC-TOPSIS method [16] is used to determine the weight of secondary indicators. The investment environment assessment of comprehensive land, sea and air coordination needs to assign weight to the water system, land system and air space system. The weight is according to the proportion of the foreign trade in goods through three transportations of sea, land and air 0.75, 0.13 and 0.12 (Table 1). GC-TOPSIS method is as follows

$$\frac{\left[\mathrm{x}_{ij} - \min\left(x_{ijj}\right)\right]}{\left[\max\left(x_{ij}\right) - \min\left(x_{ij}\right)\right]} \tag{1}$$

$$\frac{\left[\max\left(x_{ij}\right) - x_{ij}\right]}{\left[\max\left(x_{ij}\right) - \min\left(x_{ij}\right)\right]} \tag{2}$$

$$\mathrm{e}_j = -\ln m \sum\nolimits_{i=1}^{m} \left(y_{ij} \times \ln y_{ij}\right) \tag{3}$$

$$\mathrm{w}_j = \frac{d_j}{\sum_{i=1}^{m} d_j}, \tag{4}$$

where, w_j denotes the index weight, d_j denotes the redundancy, x_{ij} is indicator.

Table 1. Main issues and challenges of the territories of new industrial development.

Destination layer	Elements layer	Index parameter	Weight
Water system (0.75)	Resource Utilization (0.5)	Sea area per capita	0.08359
		Coastline length per capita	0.08345
		Length of wharf	0.08349
		Mariculture area	0.08341
		Shoreline economic density	0.08332
		Inland waterway density	0.08341
	Industrial Economy (0.25)	People employed in water transport industry	0.08308
		Gross ocean product per 10000 people	0.08322
		Gross marine product (% of GDP)	0.08294
	Social Development (0.17)	Port cargo through	0.08327
		Passenger capacity	0.08353
	Investment convenience (0.08)	Water accessibility	0.08327
Land system (0.13)	Resource Utilization (0.36)	Cultivated land area per 10000 people	0.07142
		Total Energy production	0.07133
		Economic density	0.07151
		Road density	0.07139
		Railway density	0.07147
	Industrial Economy (0.21)	Fixed investment	0.07128
		Total volume of trade	0.0714
		GDP growth rate	0.07164
	Social Development (0.36)	Per capita disposable income of urban households	0.07139
		People employed in road transport industry	0.07147
		Cultural service enterprises	0.07126
		Students in general higher education	0.07138
		Cargo capacity	0.07138
	Investment convenience (0.07)	Land accessibility	0.07166
Air system (0.12)	Resource Utilization (0.29)	Airport density	0.14297

(*continued*)

Table 1. *(continued)*

Destination layer	Elements layer	Index parameter	Weight
		Routes	0.14279
	Industrial Economy (0,43)	Persons employed in air transport	0.14270
		Passenger throughput	0.14291
		Cargo throughput	0.14292
	Social development (0.14)	Added value of transportation in industry	0.14259
	Investment convenience (0.14)	Air accessibility	0.14313

3 Results and Discussion

3.1 Investment Environment Evaluation

From the provincial sea, land and air system investment environment difference. Hainan, Guangdong, Shandong, Zhejiang and Liaoning are the regions dominated by the water system for investment environment. They have rich marine resources and good marine ecology, and are the regions with the most potential for investment and cooperation in the fields of Marine system resources, environment and industry. Beijing, Shanghai, Zhejiang, Jilin, Heilongjiang and so on are the regions dominated by the land system for the investment environment. This region has a relatively strong advantage of land industry cluster, outstanding achievements in land economic construction, and a good foundation for land industry development and cooperation. There are 15 airports in Sichuan and 14 airports in Yunnan, with densities of 32,200 km^2 and 26,200 km^2 respectively. The advantage of high-quality airports can break the geographical barrier and information barriers, and the "land-air linkage" can provide a new path for the economic development of this region (see Fig. 1).

3.2 Water System

We establish an index system to evaluate the investment environment of the water system in northeast China from four dimensions: resource utilization, industrial economy, social development and investment convenience [17]. As shown in Fig. 2, the development and utilization of Marine resources are the basic conditions for investment in water systems, which determine the long-term stable operation and future development potential of the region. Liaoning has excellent seaports, abundant natural resources, complete allocation of production factors and good talent reserve. Liaoning is the province with the best investment environment in northeast China.

According to the score of secondary index of northeast China's Water system investment environment evaluation, all indicators of Liaoning province's water system are relatively developed. But, the inland waterway density is relatively backward, which is not conducive to the radiation of port regional advantages to inland areas and drives regional

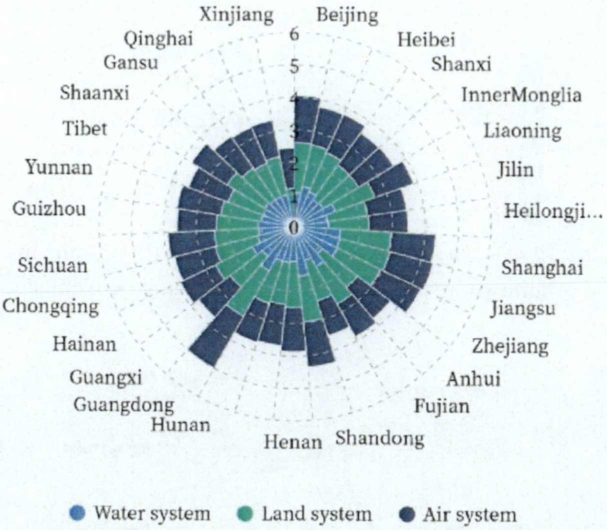

Fig. 1. Land, sea and air Investment environment between provinces in 2020.

coordinated development. The water system resource structures of Heilongjiang and Jilin provinces are highly similar. Due to the lack of coastal advantages, Heilongjiang and Jilin vigorously develop inland waterways, with inland waterway densities of 112 km/km^2 and 77 km/km^2 respectively. It will narrow the gap in Marine resources with Liaoning Province, undertake the high-quality resources attracted by the coastal areas of Liaoning Province, to realize the effective linkage of land and sea to provide important foundation support.

3.3 Land System

Land system investment is mainly reflected in the development of land economy based on geographical location, land resources, and land industrial cluster. Social development foundation and resource richness are important determinant dimensions of investment environment of land system (see Fig. 3).

In the evaluation of investment environment of land system in northeast China, there is little difference in the level of investment environment of land system in three provinces. Heilongjiang has the advantage of resources; Liaoning has the advantage of social development. Regions should make use of the linkage advantages of resources and development foundation to promote the transformation and upgrading of traditional industries, enhance the overall attractiveness of regions, and promote coordinated regional development.

According to the score of the secondary index of the land system investment environment evaluation in northeast China, the land system investment environment of the northeastern provinces has its own advantage. Heilongjiang has rich black soil resources, the largest arable land area per 10,000 people, and the land economy based on agriculture is developing strongly. Jilin has a good business environment and broad market

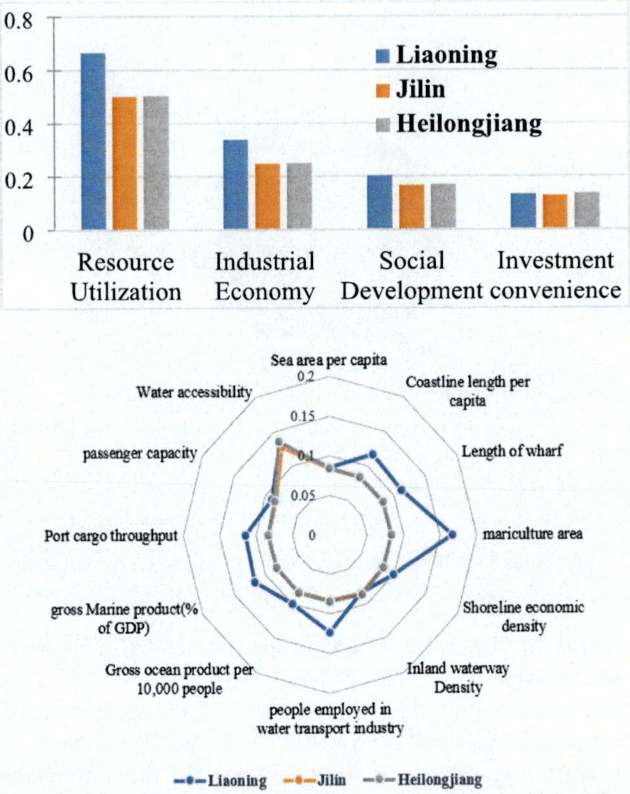

Fig. 2. Investment environment evaluation of Water system.

themes, strong economic foundation and strong enthusiasm for cooperation. Liaoning is superior to other provinces in road network density and talent training, with perfect talent structure, convenient investment and good economic development potential in land areas.

3.4 Air System

Air investment environment is mainly reflected in the combination of regional good air conditions and resource environment development around air transport, air tourism and other production activities. As shown in Fig. 4, the development of the investment environment of the airspace system in the three northeastern provinces is relatively balanced, and it has a good foundation for the development of the air economy in terms of industrial economy and accessibility.

There is little difference in the scores of the secondary indexes of the investment environment of the airspace system in northeast China, which are all at the medium level among the provinces of China. Among them, Liaoning has 524 air routes, higher than Heilongjiang and Jilin 242 and 372, with frequent flights and good air transport

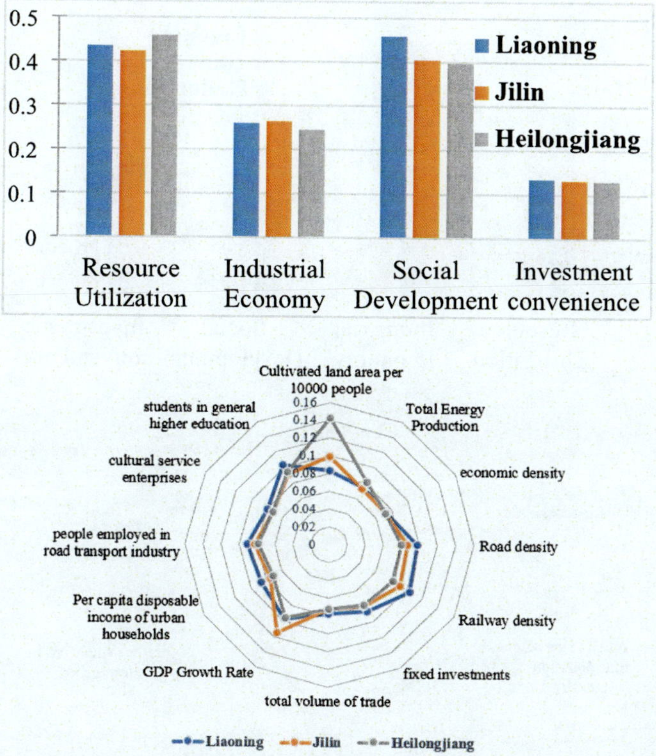

Fig. 3. Investment environment evaluation of Land system.

resources. They have wide coverage of airspace resources, the airport densities of Heilongjiang and Jilin are 3.49 km^2 and 3.13 km^2, respectively. In terms of air accessibility, Northeast China is located in the northeast of China, and its accessibility level lags behind that of China. Northeast China is bordered by many neighboring countries, and the development of airspace system construction can break the geographical barrier, which is a key potential area for promoting transnational cooperation in the future.

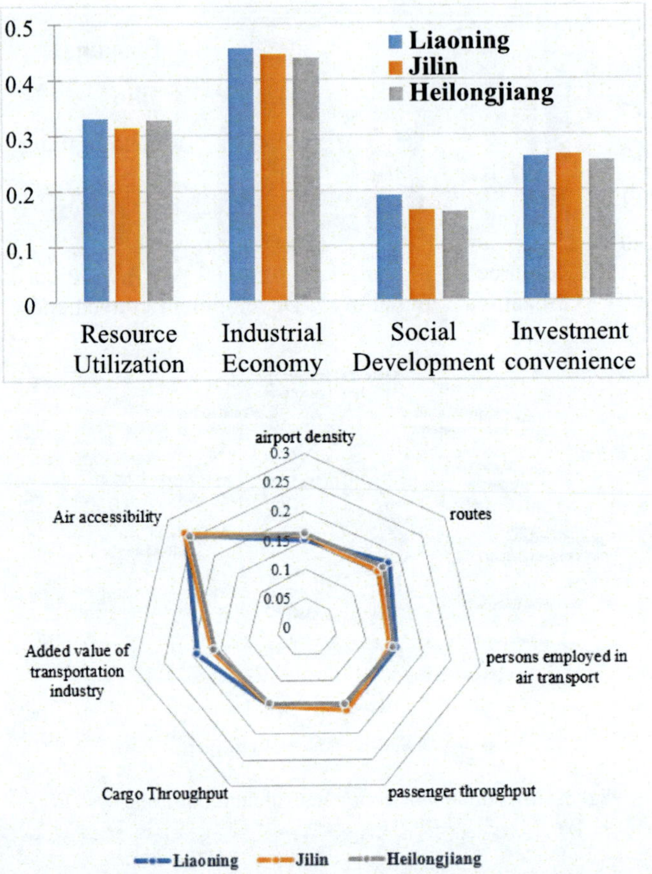

Fig. 4. Investment environment evaluation of Air system.

4 Conclusion

From the perspective of Land-Sea-Air coordination, we analyze the investment environment level and the difference between provinces in northeast China. Conclusions and recommendations as follow.

From the perspective of Land-Sea-Air coordination, the investment environment of the three Northeast provinces is in the middle level of all provinces and regions in China. Liaoning, with rich Marine resources and Marine industry foundation, is the region with the leading investment environment of Water system, ranking 8th in the investment environment evaluation of 31 provinces in China. Heilongjiang and Jilin are the regions dominated by land system investment environment, ranking 22nd and 24th respectively in China's investment environment assessment. In recent years, the work of Land-Sea-Air coordination in northeast China has achieved some initial results, but there is still a large space for improvement.

Water system investment environment of Liaoning strong development, but inland waterway construction slightly backward. Jilin and Heilongjiang, which have limited Marine resources, should further strengthen the Marine industry cooperation with Liaoning. To realize the transformation and upgrading of traditional industries, pay attention to the effective integration of land and sea resources, strive to form the development mode of land and sea linkage, and create a new situation of benign interaction between land and sea.

The investment environment of the land system in the three northeast provinces of China is relatively balanced, with prominent resource richness and social development friendliness. The industrial development pattern is characteristic. In the future, it is still necessary to avoid the unfavorable situation of being independent and make full use of the characteristic resources of different regions to promote the coordinated development of the land economy.

The three northeast provinces in China have a good foundation for the development of the air economy in terms of industrial economy and accessibility. However, because they are located in the border of China, the investment environment of air system is still in a backward position in the provinces of China.

We make a quantitative attempt to evaluate the regional investment environment from the perspective of Land-Sea-Air coordination. It is of guiding significance for the formulation of development strategies and policies in northeast China to alleviate the problems of land resource shortage and environmental deterioration and coordinate the development of sea, land and air systems.

Acknowledgements. Class A strategic leading science and technology project of Chinese Academy of Sciences (Grant number XDA20030203), Key Collaborative Research Program of the Alliance of International Science Organizations (Grant number ANSO-CR-KP-2020-02, ANSO-CR-KP-2020-06), National Natural Science Foundation of China (Grant number 32161143029, 42071282).

References

1. Zhang, X., Chen, X., Fang, Z., et al.: Investment in energy resources, natural resources and environment. Evidence from China. Resourc. Policy **76**, 102707 (2022)
2. Verbeke, A.: Multinational enterprises and the global economy. J. Int. Bus. Stud. **39**, 1236–1238 (2008)
3. Kahouli, B., Maktouf, S.: The determinants of FDI and the impact of the economic crisis on the implementation of RTAs: a static and dynamic gravity model. Int. Bus. Rev. **24**, 518–529 (2015)
4. Baade, R.A., Matheson, V.A.: An evaluation of the economic impact of national football league mega-events. Economics of the national football league. The State of the Art 2, 243–258 (2012)
5. Kang, Y.-Q., Xie, B.-C., Wang, J., et al.: Environmental assessment and investment strategy for China's manufacturing industry: a non-radial DEA based analysis. J. Clean. Prod. **175**, 501–511 (2018)
6. Özdemir, D.: The distribution of foreign direct investments in the service sector in Istanbul. Cities **19**, 249–259 (2002)

7. Deng, S., Li, Y., Chen, J.: Evaluating foreign investment environment in China: a systematic approach. Eur. J. Oper. Res. **100**, 16–26 (1997)
8. Chen, Y.S.: International visibility as determinants of foreign direct investment: an empirical study of Chinese provinces. Soc. Sci. Res. **76**, 23–39 (2018)
9. Head, K.: Inter-city competition for foreign investment: static and dynamic effects of China's incentive areas. J. Urban Econ. **40**, 38–60 (1996)
10. Han, Z., Di, Q., Zhou, L.: Discussion on the connotation and target of sea-land coordination. Marine Econ. **2**, 10–15 (2012)
11. Han, Z., Xia, K., Guo, J., et al.: Research of the level and spatial differences of Land-sea coordinate development in coastal Areas based on global-Malmquist-Luenberger index. J. Nat. Resour. **32**, 1271–1285 (2017)
12. O'Hara, I.M., Tanticharoen, M., MacRae, E., et al.: Land and sea: addressing the challenges facing inter-regional ecosystems in developing a sustainable bioeconomy. EFB Bioecon. J. **1**, 100017 (2021)
13. Zhang, G.: Research on the evolution logic of China's regional coordinated development strategy. On Econ. Problems **7**, 13 (2018)
14. Ji, X., Huang, X., Chen, Y., et al.: Comprehensive suitability evaluation of spatial development and construction land in the perspective of land-ocean coordination: a cases study of Liaoning province China. J. Natl. Resourc. **34**, 451–463 (2019)
15. Xia, K., Guo, J., Han, Z., et al.: Analysis of the scientific and technological innovation efficiency and regional differences of the land–sea coordination in China's coastal areas. Ocean Coast. Manag. **172**, 157–165 (2019)
16. Heerden, Q.V.: Accessibility, affordability, and equity in long-term spatial planning: perspectives from a developing country. Transp. Policy **120**, 104–119 (2022)
17. Cedillo-Campos, M.G.: How to measure and monitor the transportation infrastructure contribution to logistics value of supply chains? Transp. Policy **120**, 120–129 (2022)
18. Huang, Y.: The intercity railway connections in China: a comparative analysis of high-speed train and conventional train services. Transp. Policy **120**, 89–103 (2022)
19. Chakraborty, S.: TOPSIS and modified TOPSIS: a comparative analysis. Dec. Anal. J. **2**, 100021 (2022)
20. Yuan, J., Lv, Y., He, G.: Sustainable development goals for oceans and marine and coastal ecosystem-based management. Acta Ecol. Sin. **37**, 8139–8147 (2017)

New Physical and Geographical Regionalization of the South of Central Siberia

Galina I. Lysanova(✉)

Sochava Institute of Geography SB RAS, Ulan-Batorskaya str.1, 664033 Irkutsk, Russia
lysanova@irigs.irk.ru

Abstract. The article considers methodological aspects and results of physical and geographical regionalization of the Republics of Tuva, Khakassia, and southern and central regions of the Krasnoyarsk Territory. Regional geosystems of the lower levels merged into geosystems of a higher rank. This process was based on the experimental data, generalization and regional-typological interpretation of the typological landscape map of the south of Central Siberia M 1:1,000,000 compiled by the author. Based on long-term studies of geosystems, we have identified the typological composition and hierarchy of geomers, established the factors and basic patterns of differentiation of the landscape structure, identified the geomers that are the cores of the corresponding geochoras, identified the hierarchy and justified the boundaries of geochoras. The regionalization scheme was constructed by synthesizing a grid of contours of a typological map of geosystems. At the lowest level of integration, the divisions of facies groups displayed on the typological map were combined into territorial integrity of the rank of macrogeochora according to the principle of neighborhood and proximity of environmental conditions. The main criteria for the synthesis were the relative unity of the geological and geomorphological structure and genesis of the territory and the uniformity of bioclimatic indicators or local variants of the structure of the altitude zone. Neighboring macrogeochoras differ in the composition of geomers, having completely different sets of geomes. Further integration of geosystems of lower levels into regional geosystems of higher rank was carried out on the basis of zonal or altitude-zone criteria.

Keywords: physical and geographical regionalization · south of Central Siberia · geosystems · geomers · geochoras

1 Introduction

The study of the structural and geographical organization and anthropogenic transformation of geosystems is a necessary stage and basis for assessing the current state of natural resource potential and planning ways of sustainable territorial development of the regions in Siberia. Differentiation analysis, mapping, systematization, classification and zoning of geosystems play an important role in structuring and receiving information of the territory. The results of dynamic and especially evolutionary changes in natural geosystems in the process of natural development eventually lead to modification and

I. N. Vladimirov et al. (Eds.): RERSDNA 2022, SPEES, pp. 105–113, 2023.
https://doi.org/10.1007/978-3-031-28978-1_12

even restructuring of their structure on the one hand. While the composition, properties and functioning of their components change and influence the configuration, mutual arrangement and patterns of relationship of geosystems. On the other hand, approaches, methods and research tools are being improved in landscape studies. Therefore, information about the physical and geographical regionalization of specific territories, necessary for planning sustainable use of natural resources, needs updating constantly.

Physical and geographical regionalization is defined as the division of landscapes into actually existing subordinate territorial systems of regional dimension [1]. Physical and geographical regionalization is a special kind of landscape systematics [2]; large territorial formations with a complex internal structure are the object of study [3]. According to N.A. Gvozdetskii, "physical and geographical regionalization is the process of identifying objectively existing territorial physical and geographical units in nature" [4, p. 10], i.e. ideally, it should be unified and independent of the goals of the study. Other geographers shared a similar opinion [5, 6]. Based on V.B. Sochava's study [7], it can be argued that physical and geographical regionalization, or classification of geochoras is a system of territorial division premised on the identification of the hierarchy of natural regions, and its tasks are reduced to the establishment of territorial division by natural characteristics. While the best way of zoning is to draw the boundaries of regions based on landscape maps and further to combine (synthesize) geosystems of lower levels into regional geosystems ranked higher [2, 4, 7]. Regional differences in the structure of the geosystem are established based on landscape cartographic analysis of the area [3, 8 and 9].

2 Materials and Methods

This article deals with the results of geosystem study in the south of Central Siberia. The author was guided by the principles and methods of mapping generally accepted in Soviet and Russian landscape studies [1–8, 10], as well as similar approaches of European and Latin American researchers [11, 12].

The research area is characterized by a wide variety of natural conditions, which led to a complex differentiation of the landscape envelope. The interpenetration of West Siberian, Central Siberian, South Siberian and Central Asian geosystems is among its main features.

There are many publications dealing with the zoning of the research region but they all are not detailed enough. Only general schemes are outlined, but there is neither a justification for drawing boundaries, nor a sufficient description of the selected natural regions [13–16]. All this is due to the subjectivity of approaches to regionalization, since this problem was reduced to establishing the boundaries of regions of different ranks (physical and geographical region boundaries), and various combinations of partial boundaries of geosystem components were used as source material.

The schemes of physical and geographical regionalization differ not only in the position of the boundaries, but also in the set and number of taxa. Thus, some authors consider the physical and geographical country as the highest unit of regionalization, while others consider the physical and geographical region. The most complex regionalization scheme proposed by Yu.P. Parmuzin, M.V. Kirillov, Yu.A. Shcherbakov [13] includes a hierarchy of taxa: physiographic country - zone - province - subzone - district. In some cases, the country is divided into regions. Many authors single out provinces within countries.

At different times, the authors compiled multi-scale maps of geosystems of individual parts of the Upper and Middle Yenisei basin (Minusinsk and Tuva basins and their mountain framing) [17–19]. The research results, generalization and regional-typological interpretation made it possible to proceed to the creation of a unified typological landscape map of the south of Central Siberia. 1:1 000 000 [20, 21], which was the basis for the scheme of physical and geographical regionalization.

We used medium-scale landscape maps, schemes of physical and geographical regionalization of the Nazarov and Minusinsk basins [17, 22], maps of administrative districts (soil, vegetation, etc.), synthesized satellite images of Landsat 7 ETM, and topographic maps. In field studies, profiling was used in combination with work on key sites selected in areas with the most complex spatial structure of geosystems. The results of the data processing of satellite images were transferred to the reference cartographic base, which was a preliminary layout of the map.

The classification of geosystems and a landscape legend are based on a system-hierarchical approach to identifying the subordination of landscape taxa and an evolutionary-dynamic interpretation of mapped units. We proposed the structure of the map legend, taking into account the location and the typological spectrum of geosystems in the planetary system [23], and then identified a wide variety of natural complexes and mapped them on the territory. A group of facies representing a single factor-dynamic series of facies reflecting changes in the degree of hydromorphism in the study area was chosen as the lowest mapped unit of geomers. Facies groups were united into facies and geoms classes. A geom includes groups of facies that are similar in material and energy exchange, genesis, structural and dynamic features and biological productivity. A facies class serves as an intermediate stage between the facies group and the geom [23].

The complex structure of the geosystems of the territory belonging to four physical and geographical regions provided an opportunity and a challenge to take into account the positioning of mapping objects in the system of physical and geographical zoning. The binding of geomers to the corresponding geochoras is reflected in the names of geomers of high ranks, starting with the geom.

The study found more than 200 facies groups, which were combined into 42 facies classes and 13 geoms. The latter belong to three groups of geoms: North Asian goletz and taiga, North Asian forest-steppe and steppe, Central Asian mountain-tundra, tundra-steppe and steppe [20].

The main problem of adequate and correct mapping of spatial integration of geosystems is the principles of drawing the boundaries of the representatives of the lower-macrogeochoras (landscape districts) and upper – physical and geographical areas – zoning taxa. The territorial integrity of facies group units shown on the typological map according to their neighborhood and proximity of environmental conditions was a principle for the lowest level of integration. The main criteria for unification (synthesis) in this case were the relative unity of the geological and geomorphological structure and genesis of the territory, the uniformity of bioclimatic indicators or local variants of the structure of the altitude zone. Following V.V. Ryumin [24], we believe that, first, it is necessary to take into account the differences of neighboring macrogeochoras in the spectrum of geomers, in particular in the presence of a completely different set of geoms for each macrogeochora. At the upper level of the regionalization, it was decided to adhere to the scheme of V.B. Sochava and D.A. Timofeev [25] with clarifications and details proposed by Yu.M. Semenov and E.G. Suvorov [26].

3 Results and Discussion

Physical and geographical regionalization is the study and systematization of geochoras based on the hierarchy of natural regions. The cores of the chorological synthesis and hierarchical generalization of the geochoras were the previously compiled schemes of the physical and geographical regionalization of KATEK (Kansk-Achinsk Fuel-Power Complex) [22] and the Minusinsk Basin [17].

The new regionalization scheme (see Fig. 1) shows 56 macrogeochoras, which are parts of 17 physical and geographical provinces belonging to four physical and geographical regions: Ob-Irtysh (OIR), Central Siberian (CSR), South Siberian (SSR) and Central Asian (CAR) [20, 21]. Each of these regions has its own special geomers.

The northwestern part of the OIR is represented by the Middle Chulym Province (I), which includes the Chulym-Chetskaya (I1) and Verkhneketskaya (I2) southern taiga plain macrogeochoras. The province is characterized by latitudinal zonality and the predominance of aligned surfaces. This is a plain divided by small rivers and with a slight slope to the north. Here we have previously identified 14 facies groups, combined into four facies classes and two geoms (taiga and subtaiga Ob-Irtysh), where dark coniferous forests on podzolic, sod-podzolic and gray forest (often gley) soils, sedge-sphagnum swamps and mixed swampy forests dominate.

The northeastern part of the region is occupied by the Middle Yenisei Province (II) of the CSR with the Yenisei-Lower Angara foothill-plain (II3) and Biryusinsk piedmont-plain (II4) southern taiga macrogeochoras. The upland-plain relief prevails here, shallow soils with fragments of bedrock serve as the predominant soil-forming substrate, and the landscape structure is dominated by southern taiga geosystems of high-altitude plains - dark coniferous blueberry–grass-green moss forests on sod-podzolic and gray forest soils, as well as blueberry-green moss pine forests on podzols. The subtaiga landscapes are represented by upland light coniferous and mixed (pine, larch-pine on gray forest, turf and sod-podzolic soils) forests. The geosystems of the province belong to 18 facies groups, grouped into six classes and two geoms (taiga and subtaiga Middle Siberian).

A. OB-IRTYSH REGION (OIR). I. MIDDLE CHULYM PROVINCE: I1 – Chulym-Chetskaya plain south taiga, I2 – Verkhneketskaya plain south taiga.

B. CENTRAL SIBERIAN REGION (CSR). II. MIDDLE YENISEI PROVINCE: II3 - Yenisei–Nizhneangarskaya foothill-plain south taiga, II4 - Biryusinskaya foothill-plain south taiga.

V. SOUTH SIBERIAN REGION (SOUTH SIBERIAN FEDERAL DISTRICT). III. ACHINSK-KRASNOYARSK PROVINCE: III5 – Verkhnechetsko-Chulymskaya piedmont-plain subtaiga, III6 Achinsk-Bogotolskaya piedmont-plain steppe, III7 Verkhne-Kemchugskaya foothill-elevated taiga, III8 Krasnoyarsk foothill-plain subtaiga-forest-steppe; IV. THE SOUTH YENISEI RIDGE PROVINCE: IV9

Fig. 1. Physical and geographical regionalization of the south of Central Siberia. Macrogeochoras.

- South Yenisei low-mountain mountain-taiga; V. KANSK-RYBINSK PROVINCE: V10 – Usolsko-Abanskaya plain-hilly subtaiga, V11 – Kansk foothill-hollow subtaiga-forest-steppe and steppe, V12 – Kansk-floodplain foothill-elevated subtaiga, V13 – Pre-sayan foothill-elevated taiga-subtaiga; VI. VERKHNECHULYMSKAYA PROVINCE: VI14 – Nazarovskaya foothill-hollow steppe and forest-steppe, VI15 – Chulym-Yenisei foothill-hollow steppe and forest-steppe, VI16 – Shirinskaya foothill-hollow steppe; VII. EAST SAYAN PROVINCE: VII17 - Ezagashsk–Sydinskaya foothill subtaiga-forest-steppe, VII18 – Mansk-Yenisei low-mountain mountain–taiga, VII19 - Verkhnemansk-Kazyrskaya medium–low-mountain mountain-taiga, VII20 – Agul-Kizirskaya mid-mountain–alpine goletz–mountain-taiga, VII21 – Agul-Tumanshetskaya low-mountain

mountain-taiga, VII22 - Syda-Tubinsk foothill-basin subtaiga-forest-steppe, VII23 – Kiziro-Kazyr high-mountain goletz-subgoletz, VII24 – Udinsk-Okinskaya high-middle-mountain goletz-mountain taiga; VIII. KUZNETSK-SALAIR PROVINCE: VIII25 – Central Kuznetsk-Alatau middle-mountain goletz– mountain taiga, VIII26 - East Kuznetsk-Abakan foothill-low-mountain taiga-subtaiga, VIII27 – Uibat-Konsunskaya mid-mountain mountain taiga, VIII28 – Upper Abakan high–mountain goletz-subgoletz; IX. MINUSINSK PROVINCE: IX29 – North- Minusinskaya piedmont-hollow steppe, IX30 – Abakan-Yenisei piedmont-plain steppe, IX31 – Yenisei-Amyl foothill-piedmont forest-steppe; X. WEST SAYAN PROVINCE: X32 – Tubo-Kebezhskaya foothill subtaiga, X33 – Amylskaya middle-low mountain mountain-taiga, X34 – Ergak-Targak-Taiginsk high-middle mountain goletz-mountain taiga, X35 – Tashtypsk-Oiskaya low-mountain taiga-subtaiga, X36 – Abakan-Kebezhskaya middle-mountain mountain-taiga, X37 – Karlygan- Kuzunskaya highland goletz-subgoletz, X38 – Dzhebashsk-Ergakskaya high-middle-mountain goletz–mountain taiga, X39 – Usinsk-Yenisei low-mountain mountain taiga, X40 – Kurtushibinskaya high-middle-mountain goletz–mountain taiga, X41 – Alash low-mountain mountain taiga-forest-steppe, X42 – Khemchik-Uyuk middle-mountain goletz–mountain taiga; XI. EAST TUVA PROVINCE: XI43 – Khutin-Sistigkhemskaya middle-mountain mountain–taiga, XI44 - Todzhinskaya low-middle–mountain mountain-taiga–steppe, XI45 - Upper Yenisei high-middle-mountain goletz–mountain taiga, XI46 Kaakhemskaya middle-mountain mountain taiga, XI47 – Sangilenskaya high-middle-mountain goletz-mountain taiga; XII. PREKHUBSUGUL PROVINCE: XII48 – Darkhato-Khubsugul high-middle-mountain goletz–mountain taiga; XIII. ALTAI PROVINCE: XIII- 49 – Shapshalskaya high-mountain goletz-subgoletz; XIV. ALTAI-MONGOLIAN PROVINCE: XIV50 – Mongun-Taiga high-mountain goletz-subgoletz; XV. CENTRAL TUVA PROVINCE: XV51 – Turan foothill-hollow forest-steppe and steppe, XV52 – Khemchik foothill-hollow steppe and dry steppe, XV53 – Ulugkhem foothill-hollow steppe and dry steppe; XVI. TANNUOL PROVINCE: XVI54 – West-Tannuol high-middle mountain goletz-mountain taiga, XVI55 – East-Tannuol high-middle mountain goletz-mountain taiga.

CENTRAL ASIAN REGION (CAR). XVII. PROVINCE – BASIN OF LARGE LAKES: XVII56 – Ubsunur mountain-basin steppe and semi-desert.

In the south, the lowland southern taiga geosystems of the OIR and the CSR border on the piedmont and foothill geosystems of the mountainous SSR, where the main part of the territory under study is located. In the regionalization realized by the Moscow State University (MSU) and the "National Atlas of Russia", the OIR and CSR are the countries of Western and Middle Siberia, and SSR is South Siberia [4, 27, 28].

In the latter, we have identified 14 provinces with 51 macrogeochoras included, the location of which is determined not only by zoning, but also by altitude, so their internal structure is very complex. For example, the Achinsk-Krasnoyarsk physical and geographical province (III), which occupies the northwestern part of the region, includes four macrogeochoras: Verkhnechetsk-Chulymskaya piedmont-plain subtaiga (III5), Achinsk-Bogotolskaya piedmont-plain steppe (III6), Verkhnekemchugskaya foothill-elevated taiga (III7) and the Krasnoyarsk foothill-plain subtaiga-forest-steppe (III8). In general, the province is a foothill stage of mountain territories (Kuznetsk Alatau and Eastern Sayan). The upland-plain gently rugged, wavy relief prevails here. Subtaiga geosystems dominate in the northern part, taiga in the central part, and forest-steppe and steppe in the southwestern and southeastern parts, partly transformed by agriculture to a significant extent.

The most complex structure of the West Sayan physical and geographical province (X), includes 11 macrogeochoras:

Tub-Kebezhskaya foothill taiga (X32), Amyl middle and low mountain taiga (X33), Ergak-Targak-Taiginsk high and middle mountain goletz-mountain taiga (X34), Tashtypsk-Oisk low mountain taiga-subtaiga (X35), Abakan-Kebezhskaya middle-mountain and mountain-taiga (X36), Karlygan-Kuzunskaya high-mountain goletz-suboletz (X37), Dzhebashsk-Ergakskaya high and middle-mountain goletz-mountain taiga (X38), Usinsk-Yenisei low-mountain mountain taiga (X39), Kurtushibinskaya high and middle-mountain goletz-mountain taiga (X40), Alash low-mountain mountain taiga-forest-steppe (X41) and Khemchik-Uyukskaya middle mountain goletz-mountain taiga (X42). The province is a complex mountain system with deep river valleys and inter-mountain basins. This mountain range, located between the Minusinsk basin in the northwest and the Tuva basins in the southeast, stretches widely to the northeast from the upper reaches of the Abakan River in the southwest to the upper reaches of the Kazyra and Uda rivers in the east, where it joins the Eastern Sayan. The landscape structure is represented by contrasting geomers: from goletz-suboletz and mountain-taiga in the mountains to forest-steppe in intermountain hollows.

Along with the taiga and goletz in the SSR, large areas are occupied by basin–forest-steppe, steppe and dry-steppe geosystems. For example, the Minusinsk province (IX) includes the North Minusinsk foothill-hollow steppe (IX29), the Abakan-Yenisei foothill plain steppe (IX30) and the Yenisei-Amyl foothill forest-steppe (IX31) macrogeochoras. Its mountain framing (Western and Eastern Sayans, Kuznetsk Alatau and Abakan ridge) have a huge impact on all processes occurring in forest-steppe and steppe macrogeochoras, and most of the territory is occupied by agricultural lands in place of meadow steppes and settled meadows on meadow-chernozem soils.

The geosystems of the Preubsunur mountain-basin steppe and semi-desert macrogeochora (XVII56) of the Basin of the Great Lakes (XVII) province, which we first attributed to the CAR, stand out sharply in the southern part of the territory against the background of geosystems of the Sangilenskii high-mountain and middle-mountain goletz-taiga macrogeochora (XI47) of the East Tuva physical and geographical province (XI), the Mongun-Taiginsk high-mountain goletz-suboletz macrogeochora (XIV50) of the Altai-Mongolian province (XIV) and the West Tannuol high and middle mountain goletz-taiga macrogeochora (XVI54) of the Province of Tannuol (XVI). It is characterized by subalpine tundra-steppe geosystems on mountain tundra, mountain-steppe coarse-humus soils and dry-steppe on light chestnut, sometimes saline, and brown semi-desert soils. In the regionalization of the MSU and the "National Atlas of Russia", the territory belongs to the Tuva mountain region of the Altai–Sayan country [4, 27 and 28].

4 Conclusion

Thus, using landscape cartographic analysis, we identified the typological composition and hierarchy of geomers, established the factors and basic patterns of differentiation of the landscape structure, identified the geomers serving as the cores of certain geochoras, identified the hierarchy and substantiated the boundaries of the existing natural regions of the south of Central Siberia by natural features. Therefore, we developed a classification of geochoras, which, based on the doctrine of V.B. Sochava's geosystems [7], is the task of physical and geographical regionalization. By synthesizing the contours of the previously compiled geomer map (typological map of geosystems) at a scale of 1:1,000,000, a new geochora map (scheme of physical and geographical regionalization) has been developed. It presents 56 macrogeochoras, which are parts of 17 physical and geographical provinces belonging to four physical and geographical regions: Ob-Irtysh (OIR), Central Siberian (CSR), South Siberian (SSR) and Central Asian (CAR).

The drawing of the regional borders was preceded by the unification of neighboring enclosures and partial generalization of the typological landscape map with the binding of the combined contours to the morphostructures of the relief. For the lowest level of integration, the main principle of geochora identification was the unification of neighboring facies groups with similar ecological conditions shown on the typological map into territorial integrity of the macrogeochora rank. The main criteria for synthesis in this case were the relative unity of the geological and geomorphological structure and genesis of the territory, the uniformity of bioclimatic indicators or local variants of the structure of the altitude zone. In the landscape sense, this meant that neighboring macrogeochoras differ in the composition of geomers, having completely different sets of geoms. Further unification (synthesis) of geosystems of lower levels into regional geosystems of a higher one was carried out based on zonal or altitude-belt criteria.

References

1. Sochava, V.B.: Principles of physical and geographical regionalization. In: Questions of Geography. Collection of articles for the 18th Internships Geography Congress, pp. 356–366. Moscow-Leningrad (1956)
2. Isachenko, A.G.: Basic principles of physical and geographical regionalization and issues of constructing a taxonomic system of units. Uchenye zapiski Leningradskogo gosudarstvennogo universiteta im. A.A. Zhdanova. Ser. Geograph. Sci. **8**(317), 21–53 (1962)
3. Isachenko, A.G.: Theory and Methodology of Geographical Science. Publishing Center Academy, Moscow (2004)
4. Gvozdetsky, N.A., Zhuchkova, V.K., Zvonkova, T.V.: Physical and Geographical Regionalization of the USSR: Characteristics of Regional Units. Publishing House of Moscow University (1968)
5. Mikhailov, N.I.: Physical and Geographical Regionalization. Publishing House of Moscow University (1985)
6. Mikheev, V.S.: Landscape Synthesis of Geographical Knowledge. Nauka, Novosibirsk (2001)
7. Sochava, V.B.: Introduction to the Doctrine of Geosystems. Nauka, Novosibirsk (1978)
8. Isachenko, A.G.: Landscape science and physical and geographical regionalization. Vysshaya Shkola (1991)
9. Suvorov, E.G., Semenov, Yu.M., Kuzmenko, E.I.: Landscapes. Geography of Siberia at the beginning of the 21st century, vol. 2. Nature. GEO, Novosibirsk (2015)
10. Suvorov, E.G., Semenov, Y., Novitskaya, N.I.: Landscape assessment map of the Asian part of Russia: principles and methodological aspects of compilation. Geogr. Nat. Resour. **4**(30), 5–10 (2009)

11. Bastian, O., Beierkuhnlein, C., Klink, H.-J.: Development and Perspectives of Landscape Ecology. Kluwer Academic Publishers, Dordrecht (2002)
12. García, A.E., Miravet, B.L., Salinas, E.: The cartography of the landscapes with systems of geographical information, like base for the diagnosis geoecology of the hydrological basin of the Rio Ariguanabo Cuba. Search Natl. Assoc. Postgrad. Investigat. Geograp. **15**(27), 169–194 (2019)
13. Parmuzin, Yu.P., Kirillov, M.V., Shcherbakov, Yu.A.: Some results of the physical and geographical regionalization of Central Siberia and the Krasnoyarsk Territory. Questions of Geography. Mysl, Moscow (1961)
14. Parmuzin, Yu.P., Kirillov, M.V., Shcherbakov, Yu.A.: Physical and Geographical Regionalization of the Krasnoyarsk Territory. Materials on the Physical-Geographical Regionalization of the USSR. Publishing House of Moscow University, Moscow (1964)
15. Likhanov, B.N.: Natural Regionalization. Central Siberia. Nauka, Moscow (1964)
16. Nikolaev, V.A., Kunitsa, M.N., Mikhailov, N.I.: New materials on the physical and geographical regionalization of the USSR. Natural and agricultural regionalization of the USSR. Publishing House of Moscow University, Moscow (1983)
17. Lysanova, G.I.: Landscape Analysis of the Agro-Natural Potential of Geosystems. Publishing House of IG SO RAN, Irkutsk (2001)
18. Lysanova, G.I., Semenov, Y., Shekhovtsov, A.I.: Geosystems of the Republic of Tyva. Geogr. Nat. Resour. **3**(34), 181–184 (2013)
19. Lysanova, G.I., Semenov, Y., Sorokovoi, A.A.: Landscape mapping of the Republic of Khakassia. Geodesy Cartography **12**(918), 16–23 (2016)
20. Lysanova, G.I., Semenov, Y., Sorokovoi, A.A.: Geosystems of the south of Central Siberia: methods and results of mapping. Geodesy Cartography **4**(970), 35–44 (2021)
21. Lysanova, G.I., Semenov, Yu.M., Sorokovoi, A.A.: Mapping of geosystems in the south of the Yenisei Siberia for environmental assessment. IOP Conf. Ser.: Earth Environ. Sci. **381** (2019)
22. Semenov, Yu.M. Landscape-Geochemical Synthesis and Organization of Geosystems. Nauka, Novosibirsk (1991)
23. Mikheev, V.S., Ryashin, V.A.: Principles and methodology for compiling a map of the landscapes of Transbaikalia. Problems of thematic mapping. - Irkutsk: Ed. IGS i DV SO AN SSSR, pp. 183–192 (1970)
24. Ryumin, V.V.: Dynamics and Evolution of South Siberian Geosystems, Nauka, Novosibirsk (1988)
25. Sochava, V.B., Timofeev, D.A.: Physical and geographical regions of North Asia. // Dokl. of the Institute of Geography of Siberia and the Far East 19, pp. 3–19 (1968)
26. Goudie, A., Viles, H.: Landscapes. In: Landscapes and Landforms of Namibia. WGL, pp. 3–25. Springer, Dordrecht (2015). https://doi.org/10.1007/978-94-017-8020-9_1
27. Map of the physical and geographical regionalization of the USSR: For higher educational institutions. Map M 1: 8,000,000. Ed. Gvozdetsky and Samoilova, G.S. GUGK, Moscow (1986)
28. Physical and geographical regionalization. National Atlas of Russia, vol. 2. https://nationalatlas.ru/tom2/402-403.html. Accessed 1 Sept 2022

Dynamic Monitoring of Ecological Environment Quality Based on Landsat-7 SLC-OFF ETM Image

Jun Ma, Khromykh Vadim Valerievich(✉), and Mingkai Sun

National Research Tomsk State University, Lenin Ave 36, Tomsk 634050, Russia
majun@stud.tsu.ru, khromykh_vadim@mail.ru

Abstract. Using remote sensing technology to evaluate the ecological environment quality is of great significance to the sustainable development of the monitoring area. Tomsk district of the Tomsk region was taken as the study area, and Landsat-7 SLC-OFF images were selected as the source data to study the restoration of data "missing" areas such as strip missing area, cloud cover area, and shadow area. The remote sensing ecological index (RSEI) integrated with four indicators of greenness, humidity, heat, and dryness were constructed using principal component analysis to study the spatiotemporal dynamics of the Tomsk ecological environment. The results show that the RSEI index can better indicate the regional ecological environment. In 2005, 2009, 2014, and 2020, the average RSEI of Tomsk district is 0.77, 0.73, 0.81, and 0.72. The ecological environment quality showed a relatively stable fluctuation. From 2005 to 2020, the annual average RSEI is 0.76, and the ecological environment quality was in a good state (0.6–0.8). The ecological environment quality in 2005 is relatively lower than in other years due to temperature factors. The overall ecological environment quality of the Tomsk district is good, and the southwest corner is the best.

Keywords: Environment quality · RSEI · Principal component analysis · Change monitoring · Spatial and temporal characteristics · Tomsk region

1 Introduction

The ecological environment is a complex ecosystem related to the sustainable development of society and economy, including water resources, land resources, and biological resources that affect human survival and development [1]. The monitoring and evaluation of ecological environment quality based on 3S technology have the characteristics of comprehensive coverage and vital timeliness. It can obtain relevant information in a timely and accurate manner. The spatiotemporal data can be traced back and has high reliability. It has many practical applications in ecological environment monitoring [2–4]. The currently widely used methods in this field are mainly divided into two categories: the ecological environment evaluation method based on a single index [5–7] and the ecological evaluation method that integrates multiple indicators [8–10]. However, in the above evaluation methods of ecological environment quality, the single index evaluation

I. N. Vladimirov et al. (Eds.): RERSDNA 2022, SPEES, pp. 114–123, 2023.
https://doi.org/10.1007/978-3-031-28978-1_13

method cannot fully reflect the region's ecological environment, and the index weight in the multi-index evaluation method is very subjective. Xu et al. [11] proposed a new Remote Sensing Ecological Index (RSEI), which makes it possible to quickly assess and control the quality of the ecological environment in the study area. The index is based on Landsat data and uses principal component analysis to integrate four indicators of greenness, humidity, temperature, and dryness. It has the advantages of objectivity, multiple indicators, visualization, and high precision. Such an objective evaluation is widely used in evaluating the quality of the ecological environment.

Tomsk is a famous educational science center in Siberia, which is regarded as an exceptional techno-innovative zone because of its advantages in science and technology. However, there are a number of problems in the ecological environment, such as the world's largest nuclear production complex, pollution of the oil refining industry, as well as problems of land and water pollution, which threaten the general and local ecological environment of Tomsk. Therefore, it is of great significance to monitor and evaluate the ecological environment quality using the RSEI model combined with remote sensing inversion technology to promote the sustainable development of the social economy in the region, which can provide the scientific basis for the ecological environment protection and high-quality development of Tomsk region.

2 Materials and Methods

2.1 Overview of the Study Area

The Tomsk district of the Tomsk region (see Fig. 1) is located at the junction of the West Siberian Plain and the spur of the Kuznetsk Alatau Mountains (84°00′E—86°14′E, 55°57′N—57°17′N). The study area is 10993.2 km^2. The Tom River, which is the largest tributary of the Ob River in the southeast of Western Siberia, divides it into two parts. The Tomsk district has a temperate continental climate. Seasonal changes occur quickly. Summer is short, winter is harsh and long, and annual rainfall is 517 mm. Forest cover is about 75% of the total area of the Tomsk district. Conifer forest mainly consists of pine, cedar, spruce and fir. Deciduous trees are mainly birch and aspen. Soils are mostly podzolic and sod-podzolic. The soil cover of the floodplain is dynamic in time and is complex in structure. The following types of alluvial soils have become widespread: sod stratified, sod, sod-gley, silty-gley, peat-gley and peat [12].

2.2 Data Sources

The primary data used in the study are mainly Landsat-7 ETM SLC-OFF images with missing bands and Landsat-7 ETM SLC-ON images with complete data. The data comes from the US Geological Survey, including four periods in 2005, 2009, 2014, and 2020. The months are concentrated from June to September, and the image quality is good. In order to ensure the reliability of the research results, the cloud cover of the selected remote sensing images is less than 7%. After vector clipping, there is no cloud and cloud shadow coverage over the study area (Table 1). Since the study area needs two images to be fully covered, it is necessary to perform mosaic and crop processing on the atmospherically corrected images. The remaining indicators of the RSEI model are obtained from quantitative inversion of remote sensing image data, and each component maintains the exact spatial resolution.

Fig. 1. Research area.

Table 1. Remote sensing image data.

Data	Spatial resolution	Scan line corrector	Image to be gap Filled (Cloud cover, %)	Image to fill gaps (Cloud cover, %)
Landsat ETM+	30 m	SLC-ON	2001-08-06 (2%)	
		SLC-OFF	2005-08-17 (3%)	2004-07-29 (6%)
			2009-06-09 (0%)	2008-08-09 (0%)
			2014-06-23 (1%)	2013-07-22 (7%)
			2020-09-15 (0%)	2019-08-24 (7%)
		SLC-ON	2001-08-06 (0%)	
		SLC-OFF	2005-08-17 (2%)	2004-08-14 (1%)
			2009-06-09 (0%)	2008-08-09 (0%)
			2014-06-23 (1%)	2013-07-22 (2%)
			2020-08-10 (0%)	2019-08-24 (5%)

2.3 Research Methods

Landsat-7 ETM SLC-OFF Data Processing. In the early stage of SLC permanent fault, Landsat-7 ETM + (SLC-ON) images can be used for strip repair before SLC failure. However, surface features change significantly with time, and SLC-ON images are challenging to use for strip repair of SLC-OFF images with long time intervals [13]. In addition, the strip positions of SLC-OFF images in the same orbit and similar periods are slightly different. Therefore, one or more SLC-OFF images with different phases can be used for strip repair. This study preliminarily evaluated the feasibility and application potential of Landsat-7 ETM + SLC-OFF image in environment quality monitoring and evaluation and decided to use the local histogram matching method [14–16].

Eco-Environmental Quality Assessment Model Based on RSEI. The construction of the RSEI model needs to couple four ecological evaluation indexes, which are directly related to the quality of the ecological environment, representing four natural ecological factors, namely, greenness (NDVI) [17], humidity (WET) [18], heat (LST) [19], and dryness (NDBSI) [20]. So, the first thing to do is to build a single component. Secondly, perform principal component analysis on a single indicator, and finally, subtract the initial RSEI state value from 1 and normalize it [21]. The final RSEI value is the result of the ecological environment quality of the study area.

Construction of Comprehensive Indicators of RSEI Model. The RSEI model is constructed based on the four indicators of greenness, humidity, heat, and dryness that humans can intuitively perceive the ecological conditions. Since the dimensions of the above four indicators are not uniform, the direct principal component analysis will affect the final results. Therefore, it is necessary to normalize each indicator so that its value is in the range of [0, 1]. The normalization formula is:

$$NI_i = \frac{I_i - I_{min}}{I_{max} - I_{min}} \tag{1}$$

where NI_i is the normalized value of year i, I_i is the index value of year i, I_{min} is the minimum value of year i, and I_{max} is the maximum value of year i.

It should be noted that the model is susceptible to water body effects, so in the study, the Modified Normalized Difference Water Index (MNDWI) [22] was used to mask the rivers and large lakes within the study area. Perform principal component analysis on the four normalized indicators, and the first principal component PC1, with the most considerable retained variance, is the RSEI Formula's initial value (13). According to the RSEI research results (the normalized value of the initial RSEI Formula (14), the RSEI is obtained by further processing, and the value range is between 0 and 1.

$$RSEI_0 = 1 - \{PC1[f(NDVI, WET, NDBSI, LST)]\} \tag{2}$$

$$RSEI = \frac{RSEI_0 - RSEI_{0\_min}}{RSEI_{0\_max} - RSEI_{0\_min}} \tag{3}$$

where $RSEI_0$ is the initial value of RSEI, $PC1$ is the first principal component obtained by principal component transformation of $NDVI$, WET, LST and $NDBSI$; $RSEI_{0\_min}$ and $RSEI_{0\_max}$ are the minimum and maximum values of the initial RSEI.

Analysis Method of Ecological Environment Quality Change. In order to intuitively reflect the spatial distribution and changes in ecological environment quality in the Tomsk district, based on the existing grading standards [23], the four RSEI values were divided into five grades: Bad (0–0.2), Not good (0.2–0.4), Normal (0.4–0.6), Good (0.6–0.8) and Excellent (0.8–1) by equal interval method to analyze the changes of ecological environment quality in Tomsk district (Fig. 2).

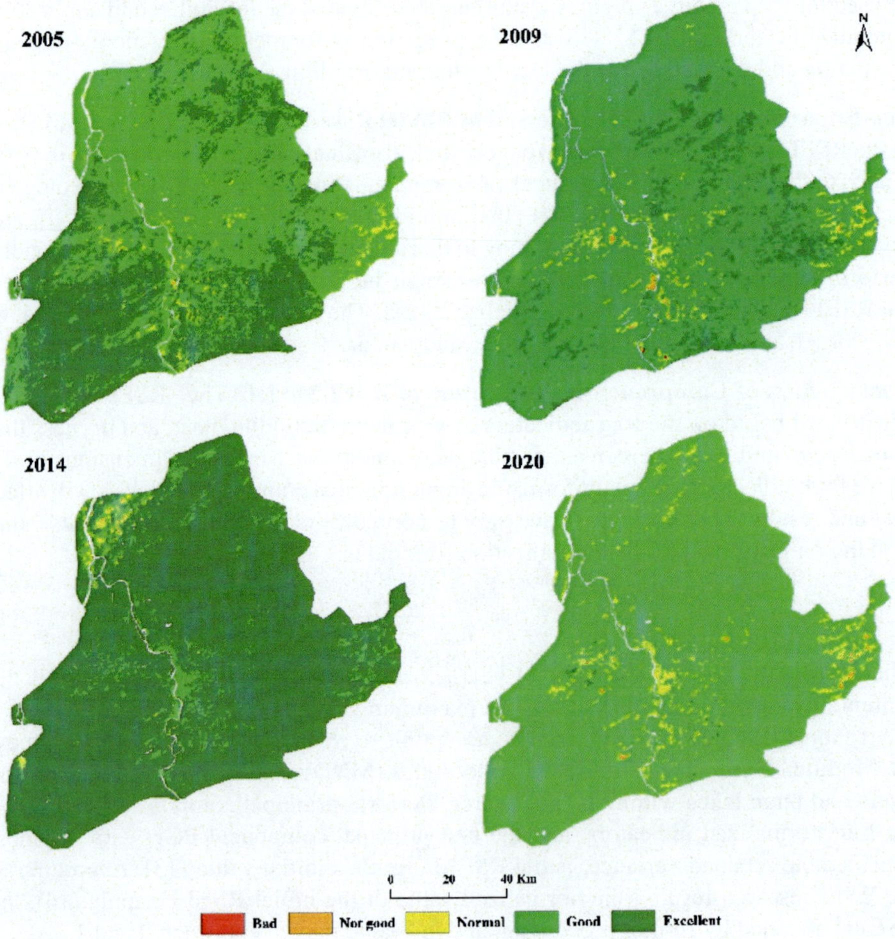

Fig. 2. The RSEI of Tomsk region.

3 Results and Analysis

3.1 Analysis of Ecological Environment Quality Index in Tomsk District

The principal component analysis was carried out on the growing season greenness (NDVI), humidity (WET), heat (LST), and dryness (NDBSI) of 2005, 2009, 2014, and 2020 in the Tomsk district. The results are shown in Table 2. Among the four indicators, the load of greenness and humidity was positive for many years, indicating that the two positively affected RSEI. The average load values of the two were 0.6875 and 0.4775, respectively, indicating that the contribution of greenness to RSEI was higher than that of humidity. The multi-year average load values of heat and dryness are −0.4725 and -0.0975, both negative, indicating that they hurt RSEI, and dryness has little contribution to RSEI. The contribution rates of the four indicators on the first principal component (PC1) reached the highest of 80.33% (2005), and the lowest was 67.41% (2014).

Table 2. The results of the first principal component analysis of the four indicators.

Year	PC1				Eigenvalues	Percent eigenvalues/%
	NDVI	WET	LST	NDBSI		
2005	0.66	0.74	−0.17	−0.03	0.01	80.33%
2009	0.65	0.41	−0.63	−0.10	0.01	76.93%
2014	0.77	0.26	−0.55	−0.19	0.01	67.41%
2020	0.67	0.50	−0.54	−0.07	0.01	78.48%
Mean	0.69	0.48	−0.47	−0.10	0.01	75.79%

3.2 Temporal and Spatial Changes of Ecological Environment Quality in the Tomsk Region

Using the RSEI model and the calculated values of four component models (see Fig. 3), the change process of eco-environmental quality in the Tomsk region can be analyzed. On the time scale, the eco-environmental quality in Tomsk region from 2005 to 2020 showed a trend of "decline-rise-decline-rise" and fluctuated within the general level. Due to the increase in rainfall and humidity in 2020, the ecological environment quality declined, which was 0.09% lower than in 2014. The results are similar to the conclusions reached by Katashova et al. [29]. on the basis of literature analysis, who chose three methods to evaluate the urban ecological status: interpolation, ecological mapping, and expert evaluation to study the urban ecological environment in the Tomsk region. The average change of each component index (after normalization) in the Tomsk region from 2005 to 2020: the average NDVI of the greenness index from 2005 to 2020 showed a downward trend and fluctuated, and the average NDVI in 2005 was as high as 0.91, indicating that the vegetation cover in the region was good. The humidity index WET showed a trend of 'rising - falling – rising', and the lowest value of WET was 0.51 in 2014. Dryness index

NDBSI overall upward trend reached the highest value of 0.42 in 2020; the heat index LST generally shows an upward trend and falls back in 2020. Overall, the change in ecological environment quality in the Tomsk region has fluctuation, but the fluctuation range is still in good condition, and the ecological environment quality is good.

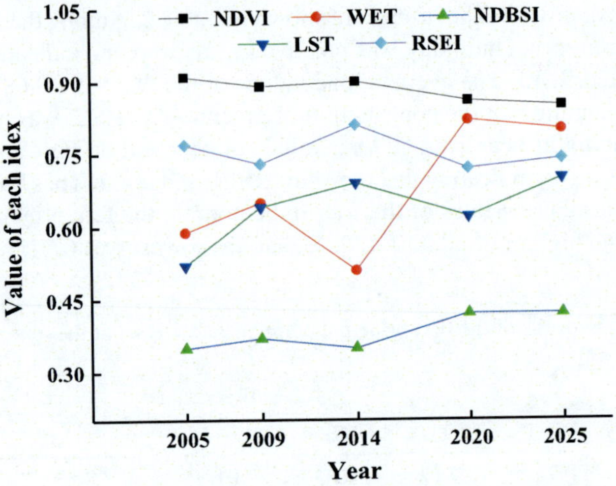

Fig. 3. Single indicator and RSEI average during 2005–2025.

Based on the above research results, the trend diagram takes five years as the step, using the average value from 2005 to 2020 and the TREND function in EXCEL to predict the ecological environment quality of the Tomsk region in 2025 of every single indicator and RSEI is obtained.

3.3 Applicability Analysis of Remote Sensing Ecological Index

In order to analyze the relationship between each index parameter and RSEI each year, the stepwise regression analysis was carried out with RSEI as the dependent variable and NDVI, WET, LST, and NDBSI as independent variables, and the relationship model was established. In this paper, 1 km × 1 km grid through the whole image sampling method, each image acquisition 1000 samples, enough samples can avoid the uncertainty of the results caused by a small number of samples. The following are regression models for four years (all models passed the 1% test of visibility):

$$\begin{aligned} \mathrm{RSEI}_{2005} = 0.521x_1 + 0.143x_2 - 0.227x_3 - \\ 0.539x_4 + 0.517\left(R^2 = 1.000\right) \end{aligned} \tag{4}$$

$$\begin{aligned} \mathrm{RSEI}_{2009} = 0.548x_1 + 0.241x_2 - 0.266x_3 - \\ 0.528x_4 + 0.458\left(R^2 = 1.000\right) \end{aligned} \tag{5}$$

$$\begin{aligned} \mathrm{RSEI}_{2014} = 0.772x_1 + 0.439x_2 - 0.591x_3 - \\ 0.295x_4 + 0.613\left(R^2 = 0.999\right) \end{aligned} \tag{6}$$

$$\text{RSEI}_{2020} = 0.508x_1 + 0.181x_2 - 0.178x_3 - 0.494x_4 + 0.453(R^2 = 1.000) \quad (7)$$

where x_1, x_2, x_3, x_4 represent the values of NDVI, WET, LST and NDBSI, respectively.

From the absolute value of the regression coefficients in the regression model each year, it can be seen that NDVI has an enormous contribution to the remote sensing ecological index, followed by NDBSI, LST, and WET. The NDVI and WET coefficients are positive, indicating a positive effect on RSEI, and the NDBSI and LST coefficients are negative, indicating a negative effect on RSEI. In the regression model, year by year, the NDVI coefficient in the positive index increases year by year, indicating that vegetation cover is more and more critical in the improvement of environmental quality. At the same time, the influence of WET is relatively weak. Among the negative indicators, except for 2014, the absolute value of NDBSI is greater than LST in other years, indicating that NDBSI has a more obvious inhibition of ecological quality. In general, the NDVI coefficient has the most considerable weight in the regression model, indicating that the Tomsk region has more significant results in implementing policies such as the rational use of forest resources and the delineation of ecological nature reserves.

In order to verify the reliability of the regression model, taking 1000 data points in 2020 as an example, a three-dimensional scatter space distribution map of each index is constructed to examine the relationship between each index and RSEI (see Fig. 4). From the three-dimensional spatial distribution of the positive indexes NDVI and WET in the left figure, the high-value areas of RSEI are mainly distributed in the areas with high vegetation coverage and good water conditions, indicating that the increase in vegetation coverage and sufficient water have apparent effects on the improvement of ecological environment quality. The three-dimensional spatial distribution of the negative indexes NDBSI and LST in the correct figure shows that the ecological environment quality gradually decreases with the increase of NDBSI and LST, and the low-value area of RSEI is mainly concentrated in the high-density building area and the high surface temperature area. In summary, the results are consistent with the above regression model results, and the model's applicability is strong.

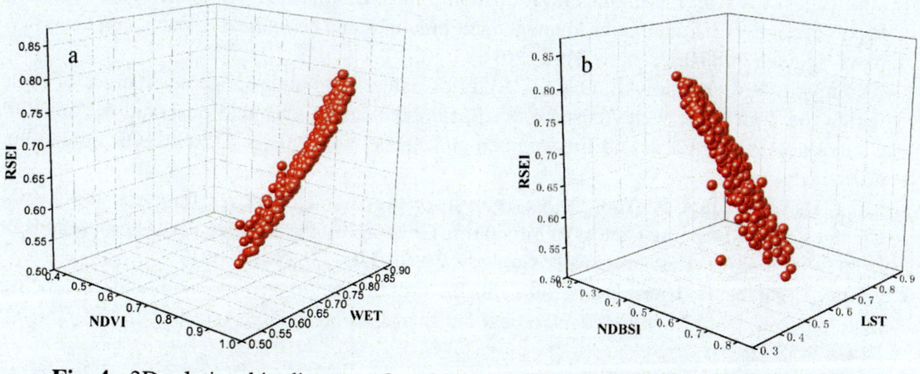

Fig. 4. 3D relationship diagram of ecological environment quality in Tomsk region.

4 Conclusions

The remote sensing ecological index RSEI established by integrating the four indicators of NDVI, Wet, LST, and NDBSI based on the principal component analysis method can objectively couple the indicators, reasonably quantify the ecological quality of the study area, and visualize the ecological quality, which can better reflect Ecological environment quality and dynamic changes.

From the RSEI regression model, NDVI and WET play a positive role in RSEI, whereas LST and NDBSI play a negative role. NDVI has the most outstanding contribution to urban ecology, but the negative impact of LST and NDBSI on urban ecology cannot be ignored. The heat index has a more significant impact on the quality of the ecological environment and is the main reason for affecting the quality of the ecological environment.

The research shows that the Tomsk region's ecological environment quality shows a fluctuating trend of "decline – rise – decline – rise" from 2005 to 2020. In 2020, the quality of the ecological environment had declined due to increased rainfall and humidity. Overall, the ecological environment quality is at a reasonable level. Predictions for 2025 according to the model, the quality of the ecological environment in the Tomsk region will be slightly lower than that in 2020, the heat component will increase, and the greenness and humidity components will decrease. Therefore, in the ecological environment construction of the Tomsk region, the protection of the ecological environment should be further strengthened, the forest coverage rate should be increased, and the management of the development of forest resources should be strengthened.

References

1. Egerton, F.N.: Ecological Economics: Energy, Environment, and Society. Juan Martinez-Alier, Klaus Schlupmann. Isis. **81**(3), 552–552 (1990)
2. Kerr, J.T., Ostrovsky, M.: From space to species: ecological applications for remote sensing. Trends Ecol. Evol. **18**(6), 299–305 (2003)
3. Khromykh, V.V., Khromykh, O.V.: Spatial structure and dynamics of Tom river floodplain landscapes based on GIS, digital elevation model and remote sensing. In: Pokrovsky, O.S. (ed.) Riparian zones: Characteristics, management practices and ecological impacts, pp. 289–309. Nova Science Publishers, New York (2016)
4. Khromykh, V.V., Khromykh, O.V.: A study of natural and anthropogenic changes of soils within the Tom River valley based on GIS, remote sensing and field observations. In: IOP Conference Series: Earth and Environmental Science, vol. 201, p. 012006. IOP Publishing (2018)
5. Liu, L.D., Li, J., Liu, C.X., Yang, X.X.: Analysis on the Characteristics of Temporal and Spatial Changes and Influencing Factors of Vegetation Coverage in the Yangtze River Economic Belt from 2000 to 2015. Res. Soil Water Conserv. **28**(6), 330–336, 347 (2021)
6. Xiong, Y., Zhang, F.: Thermal environment effects of urban human settlements and influencing factors based on multi-source data: a case study of Changsha city. Acta Geogr. Sin. **75**(11), 2443–2458 (2020)
7. Wu, X.R., Na, X.D., Zang, S.Y.: Application of temperature vegetation dryness index in the estimation of soil moisture of the Songnen Plain. Acta Ecol. Sin. **39**(12), 4432–4441 (2019)

8. Fu, B.J.: The evaluation of eco-environmental qualities in China. Chin. J. Popul. Resourc. Environ. **2**, 48–54 (1992)
9. Zhang, P., Xu, H.L., Du, Q., Ling, H.G., Zhang, P., Zhao, X.F.: Change of ecological conditions in the mainstream area of the Tarim River based on RS and GIS during the period of 1990–2010. Arid Zone Res. **34**(2), 416–422 (2017)
10. Bao, R., Qu, S., Tang, M.F., Deng, H.G.: Evaluation of ecological carrying capacity in Xilingol league based on DPSIR model. Ecol. Econ. **36**(8), 139–145 (2020)
11. Xu, H.Q.: A remote sensing index for assessment of regional ecological changes. Chin. Environ. Sci. **33**(5), 889–897 (2013)
12. Khromykh, V.V., Khromykh, V.S., Khromykh, O.V.: Features of soils in the floodplain landscapes of Siberian Rivers. In: IOP Conference Series: Earth and Environmental Science 201,012007. IOP Publishing (2018)
13. Landsat 7 SLC-off Gap-Filled Products Phase Two Methodology | U.S. Geological Survey. https://www.usgs.gov/media/files/landsat-7-slc-gap-filled-products-phase-two-methodology. Accessed 10 July 2022
14. Scaramuzza, P., Micijevic, E., Chander, G.: SLC gap-filled products phase one methodology. Landsat Technical Notes 5 (2004)
15. Li, X.W.: Landsat-7 SLC-OFF ETM Data download and its application in cyan-bacterial bloom monitoring in Lake Taihu. Admin. Techn. Environ. Monitor. **21**(03), 54–57 (2009)
16. Xu, H.Q.: A remote sensing urban ecological index and its application. Acta Ecol. Sin. **33**(24), 7853–7862 (2013)
17. Morawitz, D.F., Blewett, T.M., Cohen, A., Alberti, M.: Using NDVI to assess vegetative land cover change in central Puget Sound. Environ. Monit. Assess. **114**(1), 85–106 (2006)
18. Lambin, E.F., Ehrlich, D.: The surface temperature-vegetation index space for land cover and land-cover change analysis. Int. J. Remote Sens. **17**(3), 463–487 (1996)
19. Xu, H.Q.: A new index for delineating built-up land features in satellite imagery. Int. J. Remote Sens. **29**(14), 4269–4276 (2008)
20. Xu, H.Q.: A Study on information extraction of water body with the Modified Normalized Difference Water Index (MNDWI). Natl. Remote Sens. Bull. **9**(5), 589–595 (2005)
21. Stoms, D.M., Hargrove, W.W.: Potential NDVI as a baseline for monitoring ecosystem functioning. Int. J. Remote Sens. **21**(2), 401–407 (2000)
22. Zhang, C.Y., Yang, S.K., Dong, X.H., Zhao, C.M., Bao, H.J., Liu, J.: Research on Spatioltemporal Evolution and Influencing Factors of Ecological Environment Quality in the Upper Yangtze River Basin Based on RSEI Index. Research of Soil and Water Conservation, pp. 1–8 (2022)
23. Katashova, A.E., Nemtsova, O.A.: Monitoring the ecological state of the urban environment on the example of Tomsk and the Tomsk region (2019)

Heavy Metals Concentration Mechanisms in the Ice of the Amur River

Aleksandra F. Makhinova(✉) and Alexey N. Makhinov

Institute of Water and Ecological Problems, Far Eastern Branch, Russian Academy of Sciences, ul. Dikopoltseva, 56, 680000 Khabarovsk, Russia
mahinova@ivep.as.khb.ru

Abstract. The stratigraphic inhomogeneity of the ice layer, content of heavy metals and their distribution in ice cores have been analyzed for the first time in the channel of the lower reaches of the Amur river. The braiding of the channel promotes formation of three varieties of ice in the ice sheet: frozen (built-up), aufeis and hummocked. The component composition of the ice sheet corresponds with that of the water under the ice. A layer-by-layer investigation of ice cores provided a means of analyzing the water quality dynamics in the period of ice setting, and the degree of pollution of the ice sheet by toxic substances. The transition of water from a liquid state to a solid state promotes its supersaturation with chemical compounds. It is found that in a supersaturated solution there occurs a crystallization of crystalline hydrates and/or salts of the elements, the temperature point of the eutectic of which is reached. It is established that the Ni and Fe concentrations have a ∧-shaped distribution with maximum values in the deepwater segment of the channel. The highest contents of Cu – 22.62 and Zn – 66.20 $\mu gr/dm^3$ are observed in the upper sheet of aufeis ice, and higher concentrations of Ni – 2.91 are characteristic for the hummocked ice. Along the right bank we observed higher indicators of Pb – 5.49 $\mu g/dm^3$ and relatively low indicators of Ni – 1.72, Cu – 17.50 and Zn – 27.66 $\mu g/dm^3$ respectively. The color gamut of the core reflects selectivity and multitemporal crystallization of chemical compounds.

Keywords: Amur river · Ice · Crystalization · Heavy metals

1 Introduction

Cryochemical processes and the structure of river ice have a great influence on the water quality. However, these processes are as yet very imperfectly understood [1–4]. Little is known about the role of the turbulent flow and temperature inversion influencing the dynamic equilibrium and concentration of heavy metals in the water under the ice. It was found that the composition of river water has a profound effect on the chemical composition of ice, and in some sections of the river the technogenic influence also takes place [5]. An assessment of the water quality in the Amur river in the period of ice setting revealed that the concentration inhomogeneity of heavy metals in the ice is associated with formation conditions of the ice layer, and with the influence of urban waste water [6].

I. N. Vladimirov et al. (Eds.): RERSDNA 2022, SPEES, pp. 124–132, 2023.
https://doi.org/10.1007/978-3-031-28978-1_14

From mid-November till late April the Amur river is covered with ice. The phase transition of water to ice begins with the cooling of its upper layers to –2 °C when the heat transfer decreases considerably. Cryogenic processes intensify physicochemical transformations in a cooled water on the boundary with the ice [7]. Since the ice builds up layer by layer from below, the inhomogeneity of content of chemical elements in them, which is clearly pronounced in ice cores, indicates the differences of the conditions of their concentration [8, 9]. Crystallization of chemical compounds at the water-ice interface and their involvement in the ice phase occur selectively [10, 11]. Identification of the concentration mechanisms for chemical compounds in the ice is extremely important for assessing the quality of water under the ice, especially in the zone of influence of industrial and urban waste waters, the hydrochemical runoff of the river in t4he winter period, and migration of chemical components in the composition of ice. However, the contemporary level of knowledge remains inadequate for interpretation of many results obtained. The mechanisms concerning the crystallization of chemical compounds in the water-ice layer remain debatable. The issues related to investigations of chemical pollution of the Amur ice in industrial zones have not been resolved.

The goal of this study was to identify the main factors and conditions for concentration of heavy metals in the ice sheet of the Amur river.

2 Objects and Methods

The investigations were made in the vicinity of Khabarovsk located on the right bank, along the section of the one-branch channel of the Amur river 1200 m in width and up to 11 m in depth. The city has an advanced industrial zone, including an oil refinery, several thermal power plants, and other production facilities.

Ice and water samples were collected in February 2020 and 2021. Sampling was accompanied by measurements of the ice thickness, river depth and the length of stretches with different types of ice in the channel cross-section. The annular borer was used to drill three wells: at the left bank (48°38.547′ N, 135°04.314′ E.), in the middle of the river (48°38.026′ N, 135°04.929′ E), and at the right bank (48°37.751′ N, 135°04.966′ E). From each well we obtained a core 15 cm in diameter and described layer by layer the color and transparency of the ice, and the presence of inclusions (see Fig. 1).

Simultaneously, a water sample was collected from under the ice. With due regard for the ice stratigraphy, the core was cut up layer by layer into five parts at every 15–28 cm. The ice samples were delivered to the laboratory where they were melted, and the volume of the resulting water was measured. The water samples were analyzed in the accredited Analytical Center of the Institute of Tectonics and Geophysics FEB RAS. The weighted phase of the water sample was separated under a vacuum by means of nucle2ar filters with the pores 0.45 μm in size. In filtrates acidulated with HNO_3 to pH-2, we determined the elemental composition of the dissolved forms of metals (Co, Cu, Zn, Cr, Fe, Mn, Cd, and Pb) by inductively coupled plasma mass spectrometry (ICPMS) (Elan 9000) using the technique PNDF 14.1.2:4.135–98. Sample preparation was done by Cand. Sci. (Biol.) S.I. Levshina (IWEP FEB RAS).

Fig. 1. Extracting the core

3 Results and Discussion

3.1 The Ice Sheet of the Amur River and Ice Crystallization Mechanisms

Freezing of the water surface gives rise to ice crystals whose lattice has a tetrahedral structure. Due to the mobility of hydrogen atoms in the lattice, crystals are capable for reorientation under the effect of loads so that the natural ice differs in the degree of orientation of crystals. In the ice sheet of Amur, three varieties of ice were identified: frozen, aufeis, and hummocked. Frozen ice (uniformly crystalline) is produced along the left bank as a belt 280–300 m in width. In the main part of the channel 800 m in width there emerged the aufeis ice 20–30 cm in thickness. The stripe of hummocked ice 100–110 m in width, with numerous minor air bubbles imparting a matte shade to it, stretches along the right bank. The structure of the ice sheet is shown in Fig. 2.

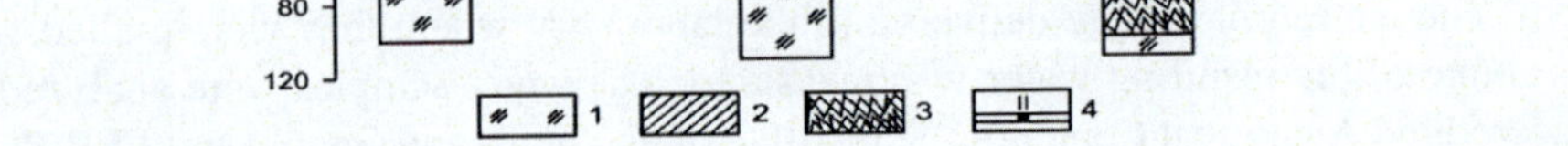

Fig. 2. Inhomogeneity of ice structure in the cross-section of the Amur river in the area of the city of Khabarovsk (February 2020). Variety of ice: 1 – frozen; 2 – aufeis; 3 – hummocked; 4 – ice and water sampling locations: I – left bank; II – middle of the river; III – right bank.

The thickness of the frozen ice throughout the winter is steadily increasing due to build-up toward the lower edge. Its maximum thickness in the middle of the channel in the area of the city of Khabarovsk reaches 125 cm, with the minimum values of 11 cm.

Build-up of the upper layers of ice is, to a greater extent, associated with the severity of climate (the duration of subzero atmospheric air temperatures) and, to a lesser extent, with the water flow velocity. However, the thickness of the ice sheet in excess of 70 cm has a warming effect in the zone of contact with the water and decreases the influence of the severity of climate. In such conditions, the current of the water flow becomes a predominant factor for ice build-up intensity. Thus, in the area of Khabarovsk, the rate of ice build-up decreases in November from 3.7 cm to 0.2 cm/day in early March.

To estimate the crystallization rate of the ice on its lower surface in the layer of cooled water we used the model [12]

$$dP/dT = \left[4/3 \cdot \pi \cdot r^3 J \cdot (1 - P)\right]/(dT/d\tau), \tag{1}$$

the solution to which was considered as a continuous function $(dP/dT) = F(P)$ on the parameter of time $[T]$ in terms of the definite $\int\int$ integral

$$dP/dT = \int F(P)dP, \tag{2}$$

where P is the ice thickness, T is time, days, 4/3 (k) is a correction coefficient, $\pi.r^3$ is the volume of the ice layer, J is the total concentration of elements of the cooled water at the contact with the ice, and τ is the volume of water of the melted ice.

Results from calculating the ice build-up rate are in good agreement with full-scale measurements which are presented in Table 1.

Table 1. Some characteristics of changes in the ice sheet of the Amur river.

Study areas	Rate of change in ice thickness, cm per day/ice thickness, cm						Air temperature and ice thickness in January	
	Nov	Dec	Jan	Feb	Mar	Apr	cp/min	min/max
Khabarovsk	3.7/32	1.3/73	0.9/99	0.6/115	0.2/117	−0.5/108	−19/−32	111/125
Komsomolsk-on-Amur	3.5/33	1.5/78	1.1/111	0.5/126	0.1/128	−0.3/122	−23/−36	125/130
Nikolaevsk-on-Amur	2.1/33	1.2/70	0.9/96	0.7/115	0.1/117	−0.2/111	−21/−32	116/119

The cooling of water below the point of its freezing promotes formation of ice crystals and an increase in concentrations of chemical compounds in the water that has remained between ice crystals. A further cooling leads to off-freezing of the whole of water, the crystallization of chemical compounds and their freezing into the ice [8]. The content of chemical compounds per unit volume of ice of its different layers primarily depends on those at the interphase boundary [5, 13].

3.2 Chemical Compounds in the Ice and Mechanisms of Ice Crystallization

Concentrations of chemical elements in different types of ice layer depend on the duration of subzero atmospheric air temperature, stratigraphic features of the ice, and on the influence of industrial sewage. On the whole, the contents of elements in the ice sheet of the hydrometric section of the channel along the direction from the right bank to the left bank obey the relationship (Mn, Zn) >> (Cu, Pb) >> Cr > Fe > Ni (Table 2).

Table 2. Mean content of metals in different types of ice, μg/dm^3.

Types of ice	Cr	Ni	Cu	Zn	Pb	Fe	Mn	n	P
Frozen	0.27	2.15	16.3	36.9	3.64	45.4	1.70	32	≤0.05
Aufeis	0.62	2.03	22.6	66.2	5.44	47.8	18.4	17	≥0.05
Hummocked	0.30	2.80	18.6	44.1	5.34	52.2	12.3	29	≥0.05

Note. n – number of core samples, P – mean value of the statistical error

Ni and Fe show a single-type distribution with their maximum content along the right bank. The indicators of Mn, Zn, Cr, Cu and Pb concentrations in the ice have a ʌ-shaped distribution in the width of the channel.

It was found that at the atmospheric air temperature below –25 °C and with the ice thickness not exceeding 80 cm, the water under the ice of the Amur develops the most favorable conditions for an intensification of cryochemical processes. An increase in the ice thickness promotes a decrease of the rate of cooling at the interface boundary, and a deceleration of crystallization processes (Table 3).

Table 3. Content of metals in different ice layers and in the water under the ice.

Ice layer, cm	Mean content of elements, μg/dm^3								
	Cr	Co	Ni	Cu	Zn	Cd	Pb	Fe	Mn
Frozen ice (left bank)									
0–20	0.13	0.02	1.72	17.50	27.66	0.02	5.49	12.62	1.16
20–40	0.34	0.02	2.29	21.02	35.60	0.02	2.77	102.77	1.15
40–60	0.96	0.08	7.02	22.74	80.15	0.05	7.91	156.81	2.58
60–80	0.19	0.02	0.79	10.56	23.59	0.02	1.45	24.50	1.17
80–100	0.12	0.03	1.45	15.98	31.85	0.02	3.68	12.59	1.57
Water	0.41	0.14	1.14	3.22	18.52	0.04	0.17	580.14	43.36
Aufeis ice (middle of the channel)									
0–23	0.62	0.09	2.03	22.62	66.20	0.03	5.44	47.55	18.40
23–46	0.17	0.03	1.13	15.17	32.86	0.01	2.57	19.19	3.81
46–69	0.19	0.02	3.03	16.30	41.74	0.02	2.64	25.63	2.18
69–92	0.16	0.02	0.91	13.96	31.09	0.03	2.57	8.90	0.96

(continued)

Table 3. (*continued*)

Ice layer, cm	Mean content of elements, μg/dm³								
	Cr	Co	Ni	Cu	Zn	Cd	Pb	Fe	Mn
92–110	0.13	0.02	0.98	13.82	27.68	0.02	3.68	–	0.74
Water	0.63	0.15	1.24	4.79	12.46	0.01	0.21	497.23	27.17
Hummocked ice (right bank)									
0–28	0.30	0.07	2.91	15.76	40.54	0.03	5.07	18.90	10.71
28–56	0.43	0.04	3.05	23.78	47.76	0.04	3.79	27.38	5.93
56–76	0.17	0.05	2.46	13.54	34.66	0.03	3.18	34.77	11.53
76–100	0.29	0.06	2.78	21.43	53.37	0.02	9.31	127.96	21.07
100–110	0.23	0.05	5.44	38.69	68.20	0.04	8.16	19.69	3.37
Water	0.58	0.15	1.14	6.62	15.14	0.03	0.14	456.49	33.98

In the zone of technogenic influence where pollution of the ice is observed, the concentrations of chemical compounds in the ice layer are in poor agreement with their content in the water under the ice [13]. Also, there occurs a concentration of chemical compounds arriving with the water due to its extrusion through cracks (see Fig. 3). When the polluted snow enters the aufeis water, it increases the contents of chemical compounds in the upper layers (0–28 cm) of all types of ice.

However, the frequency of cracks and the increased mineralization of the vertical flow manifest themselves, to a greater extent, in the 45–70 cm layer of the ice sheet (Table 3). Thus this layer develops increased concentrations of Ni, Cu, Zn, Pb and Fe. The indicators of their values can exceed one order of magnitude those in the water under the ice. Near the city however, there occurs some selectivity of the elements in different types of ice. Thus a low content of Cu – 22.62 and Zn – 66.20 μg/dm³ is observed in the upper layer of the aufeis ice, and higher concentrations of Ni – 2.91 and Zn – 40.54 μg/dm³ occur in the hummocked ice.

Along the left bank, in the upper layer of frozen ice we observed high indicators of Pb – 5.49 μg/dm³ and low indicators of Ni – 1.72; Cu – 17.50 and Zn – 27.66 μg/dm³. Pollutants affect the color gamut of the core (see Fig. 4). In the 0–28 cm layer, the ice has a greenish-dirty blue color, which is due to increased contents of zinc and copper. The dark layer of br3ownish-grey color showed high concentrations of Fe and Pb with the inclusion of dust with increased contents of Ni [14]. The concentration processes of heavy metals (Fe, Mn, Ni, Cu, Zn, and Pb) in the lower layers of hummocked ice (76–110 cm) differ from those described above. With the ice thickness of 70–80 cm, the temperature of the water under the ice varies from –1° to –2°, and the atmospheric air temperature is below –20 °C [7].

As a result, there emerge specific conditions for cryochemical processes. First, the 70–80 cm layer with maximum indicators of concentrations of elements according to the time of formation corresponds to the middle of winter when the atmospheric air temperature below –20 °C lasts for more than 20 days, and the thickness of the ice sheet is

Fig. 3. Ice cracks

Fig. 4. Multi-layered ice core

about 70 cm. The rate of ice build-up during this period (up to 80 cm) is 0.3–0.4 cm/days. Overcooling of the water under the ice first of all promotes formation of ice crystals. As a result, mineralization of water in the intercrystaline ice space increases to reach its supersaturation [11, 15]. A complete freezing-off of water is accompanied by deposition of crystals of all compounds dissolved in it (salts and/or crystalline hydrates). Second, taking into consideration that the natural waters of the Amur basin are iron-manganese waters, a further cooling of the water under the ice and ice crystallization will continue until the concentration of the water under the ice reaches the saturation point of iron and manganese as elements governing the hydrochemical runoff of the Amur river. The point of saturation temperature of an element in the water and crystallization of its compounds is said to be the temperature eutectic point or the crystalline drat point [16, 17].

The background content of soluble forms of iron and manganese, respectively, in the river exceeds 100 and 50 $\mu g/dm^3$ (Table 3). The high concentrations of ir3on compounds maintain a turbulence in the water at the contact with the ice surface, and an intensification of dissolution processes of compounds manifesting an affinity to iron, such as copper, zinc and lead, with the production of the cations Cu^{+2}, Zn^{+2}, and Pb^{+2} [15].

Considering that the crystallization of salts occurs over a certain temperature range, all components with the eutectic system higher than iron salts (eFe –1.824 °C) are involved earlier in cryochemical processes. They include, for example, copper (eCu –1.6 °C) [16, 18]. As a result of the freezing-in of salts into the ice, the eutectic points develop the building-up mass consisting of very small crystals of chemical compounds, mostly crystalline hydrates. With a rise of temperature, crystalline hydrates of salts become unfrozen first, by moving which it is possible to obtain a clean water of a high quality [18].

Results presented in Table 3 indicate high concentrations of Fe, Mn, Ni, Cu, Zn and Pb in the ice, exceeding one order of magnitude their indicators in the water under the ice. However, the temperature eutectic boundaries of different salts make it possible to explain the concentration processes of chemical components in ice formations which are sometimes impossible in terms of their content in the water under the ice (Table 2) [17, 19, 20].

Ice crystals are relatively large in size (the transverse size varying from a few fractions of a millimeter to several tens of centimeters). They are capable for reorientation under the effect of loads, which affects their metamorphization. However, the properties of ice are very poorly understood. And it is conceivable that the causes for high contents of chemical compounds inside the ice are not, to a certain extent, due to the nature of cryo-physicochemical processes as much as to the amphoteric properties of water associated with its capability for ionization at the water-ice interface.

4 Conclusions

The findings reported in this paper provide insight into the cryochemical mechanisms of concentration and crystallization of chemical compounds in the water-ice interphase. During the period of ice setting, the processes of nucleation of crystals of chemical compounds are governed by the hydrochemical composition of water and their concentration at the eutectic point. The transition of water from the liquid phase to the solid phase with the production of ice crystals promotes an increase in concentrations of chemical compounds in the intercrystalline ice space, reaching its supersaturation. Total freezing of water is accompanied by a crystalization of compounds dissolved in it (salts and crystalline hydrates).

The ice layer showed high contents of Ni, Cu, Zn, Pb and Fe relative to their concentrations in the water under the ice. Their maximum values are observed in the 70–80 cm layer, which corresponds in terms of time to mid-winter when the atmospheric air temperature below –20 °C lasts for more than 20 days. The rate of ice build-up during this period is 0.3–0.4 cm/day.

The increased concentrations of elements in the 40–60 and 80–110 cm ice layers along the right bank are due to the influence of technogenic factors. However, the nucleation of crystals of chemical compounds occurs selectively, in accordance with the cooling of water and with the temperature point of their eutectic system, which is indicated by the different relationship of elements in the layers of the ice sheet. Thus, the cryochemical processes promote concentration and crystallization of chemical compounds in the water-ice interface and a decrease in their concentrations in the water flow of the Amur. The color gamut of the cores reflects a multitemporal crystallization of chemical compounds and t4heir selectivity in this process.

Acknowledgments. This research was done with the financial support from the Russian Foundation for Basic Research under scientific project no. 19–55–80022/21.

We are grateful to Dr. Sci. (Chem.), Professor of the Faculty of Chemistry, Lomonosov Moscow State University, I.V. Perminovoiz for consultations and valuable suggestions in the preparation of this paper.

References

1. Makhinov, A.N., Kim, V.I., Matveenko, D.V.: Structure and long-term dynamics of the ice cover in the lower reaches of the Amur River. Ice Snow **58**(1), 117–126 (2018)

2. Zobrist, B., Marcolli, C., Peter, T., Koop, T.: Heterogeneous ice nucleation in aqueous solutions: the role of water activity. J. Phys. Chem. **112**, 3965–3975 (2008)
3. Wang, T., Guo, X., Fu, H., Guo, Y., Peng, X., Wu, Y., Li, J., Xia, Y.: Effects of water depth and ice thickness on ice cover blasting for ice jam flood prevention: a case study on the Heilong River, China. Water **10**(6), 700 (2018). https://doi.org/10.3390/w10060700
4. Kim V.I., Makhinov A.N.: Ice erosion of the Amur River banks. 21-st IAHR International symposium on ice, China, June 11 to 15 2012, pp. 150–159. University of Technology Press, Dalian (2012)
5. Makhinov, A.N., Kim, V.I., Makhinova, A.F., Matveenko, D.V.: The content of heavy metals in the ice of the lower reaches of the River. Ice Snow **62**(2), 251–260 (2022)
6. Ivanov, A.V.: Formation of the chemical composition of natural ice in Siberia and the FarEast. Mater. Hydrochem. Res. **67**, 67–83 (1989)
7. Ivanov, A.V.: Theory of Cryogenic and Glycogenic Hydro Chemical Processes. VINITI, Moscow (1987)
8. Petrov, E.S., Novorotsky, P.V., Lenshin, V.T.: Climate of the Khabarovsk Territory and the Jewish Autonomous Region. Dalnauka, Vladivostok - Khabarovsk (2000)
9. Ivanov, A.V.: Cryogenic metamorphization of the chemical composition of natural ice, freezing and melt waters. Dalnauka, Khabarovsk (1988)
10. Erdemir, D., Lee, A.Y., Myerson, A.S.: Nucleation of crystals from solution: classical and two-step models. Account. Chem. Res. **42**(5), 621–629 (2008)
11. Wilson, P.W., Heneghan, A.F., Haymet, D.J.A.: Ice nucleation in Nature: supercoiling point measurement and the role of heterogeneous nucleation. Cryobiology **46**, 88–98 (2003)
12. Chukin, V.V., Platonova, A.S.: Crystallization of supercoiled droplets of aqueous solutions. Natl. Techn. Sci. **4**, 231–236 (2009)
13. De Mott, P.J.: Quantitative descriptions of ice formation mechanisms of silver iodide type aerosols. Atmos. Res. **38**(1), 63–99 (1995)
14. Matoba, S., Sasaki, H., Shiraiwa, T.: Iron flux over the subarctic pacific estimated by an ice-core record from Mount Wrangell, Alaska. In: Shiraiwa, T. (ed.) Repot of Amur-Okhotsk Project, vol. 5, pp. 189–198. Research Institute for Humanity and Nature, Kyoto (2008)
15. Fukuta, N.: A study of the mechanism of contact ice nucleation. J. Atmos. Sci. **32**(8), 1597–1603 (1975)
16. Algorithm for calculating the eutectic point when using active ice sensors. https://naukarus.com/algoritm-rascheta-tochki-evtektiki-pri-ispolzovanii-aktivnyh-datchikov-gololeda. Accessed 22 June 2022
17. Eutectic points and super saturation in cooling water. https://www.activestudy.info/evtekticheskie-tochki-i-peresyshhenie-pri-oxlazhdenii. Accessed 22 June 2022
18. Shaw, R.A., Durant, A.J., Mi, Y.: Heterogeneous surface crystallization observed in undercooled water. J. Phys. Chem. **109**(20), 9865–9868 (2005)
19. Moore, E.B., Molinero, V.: Structural transformation in supercoiled water controls the crystallization rate of ice. Nature **479**, 506–508 (2011)
20. Ehre, D., Lavert, E., Lahav, M., Lubomirsky, I.: Water freezes differently on positively and negatively charged surfaces of pyroelectric materials. Science **327**, 672–675 (2010)

On the Possibility of Forecasting Abrupt Differences in Average Daily Air Temperature (a Case Study of Irkutsk)

S. V. Morozova[1](✉), E. A. Polyanskaya[1], K. E. Denisov[2], K. S. Kondakov[3], V. A. Gusev[1], and M. A. Alimpieva[1]

[1] Chernyshevskii Saratov State University, 410012 Saratov, Russia
swetwl@yandex.ru
[2] Vavilov Saratov State University of Genetics, Biotechnology and Engineering, 410003 Saratov, Russia
[3] Russian Research and Design-Technological University of Sorghum and Corn, 410050 Saratov, Russia

Abstract. This paper presents the technique for forecasting abrupt warmings and coolings during a month, based on the model developed in the Department of Meteorology and Climatology of Saratov State University. The model is built upon the principle of "floating analog" as suggested by V.F. Martazinova on the basis of the quasi-bimonthly rhythmicity of atmospheric processes substantiated by her. The Euclidean distance formula was used to determine the most typical situations in which there occur abrupt warmings and abrupt coolings. Such situations are referred to as reference situations, and when making a forecast, they serve as predictors. The identification correctness of reference fields is confirmed by synoptic analysis. It was found that winter coolings in Southern Siberia are governed by two types of processes: a radiation cooling in the central region of the winter Asian anticyclone, and advection of cold air masses along the eastern periphery of the winter Asian anticyclone. The warmings in the winter in the south of Siberia are associated with warm advection across the southern part of an extensive depression adjacent to the northern periphery of the region of high pressure located further to the south. The correctness of dates of differences in average daily air temperature according to an independent sample made up 71% for heat waves, and 69% for cold waves. Such a correctness can be considered highly successful in view of the rather rigid criterion of wave correctness and the monthly forecast lead time.

Keywords: Synoptic situation · Model process · Long-term forecast · Heat and cold waves · Analog method

1 Introduction

The ongoing warming manifesting itself everywhere is characterized not only by a rise in air temperature but also by an increase in recurrence of abnormal and unfavorable

I. N. Vladimirov et al. (Eds.): RERSDNA 2022, SPEES, pp. 133–141, 2023.
https://doi.org/10.1007/978-3-031-28978-1_15

weather conditions [1–5]: heavy showers and snowfalls, squally enhancement of wind, tornadoes, frosts and glaze-hoarfrost deposits. These phenomena are all unfavorable for the operation of weather-dependent sectors of the economy. Therefore, advance information on the expected unfavorable or dangerous phenomenon would help to considerably reduce hydrometeorological risks.

Note that the earliness of forecasting all of the aforementioned phenomena varies from a few hours to one or two days. Therefore, it is relevant to increase the forecast lead time for dangerous and unfavorable weather conditions, which is a very complex scientific problem. However, the overwhelming majority of these phenomena occur during abrupt differences in average daily air temperature; hence, this dictates the need to forecast the average daily air temperature differences with the largest possible lead time. At present there are no reliable methods of forecasting this phenomenon (an abrupt difference) with a lead time exceeding the limit of predictability of hydrodynamic models (10–14 days). Therefore, it is worthwhile and highly relevant to develop techniques of forecasting abrupt differences in men daily temperature with a lead time exceeding the limit of predictability of the atmosphere. It should be noted that a publication of specialists of the Hydrometcenter of Russia [6] points to a need to improve the quality of forecasts of meteorological extremums, and to the fact that this direction is a priority for the national weather service. Foreign publications, such as [7, 8] also indicate the need to study heat and cold waves and to develop technique of forecasting them.

The objective of this paper is to test the model of long-term forecasting of abrupt differences in average daily air temperature for the territory of Southern Siberia. The model is built upon the principle of "floating analog" as suggested by Dr. Sci. (Phys.–Math.) V.F. Martazinova on the basis of the quasi-bimonthly rhythmicity of atmospheric processes substantiated by her [9–12].

It should be noted that the model was developed in the Department of Meteorology and Climatology of Saratov State University and is being used in forecasting in the routine mode. The model permits forecasting not only abrupt differences in average daily air temperature but also period of atmospheric precipitating with a monthly lead time [13, 14]. The information base and software, created for this purpose, allow such forecasts to be made also for other regions [15–17].

2 Materials and Methods

The object for study includes abrupt differences in average daily air temperature. The most convenient and widely used method of representing abrupt differences in average daily temperature are heat waves and cold waves. Note that a generally accepted definition of the "heat wave" and "cold wave" does not exist yet [6, 18, 19].

It is thought that B.I. Sreznevskii was the first to introduce into the national scientific literature the term "cold wave" [20]. He suggested that the cold wave should imply a decrease in air temperature by 10 °C or more per day. Later, a similar method of definition was also extended to the heat wave. Usually, the heat (cold) wave is taken to imply a change in average daily air temperature from day to day with respect to the normal value [21–23]. Sometimes the values of standard deviations are used to determine the heat/cold wave, as is done, for example, in [24]. Particular percentiles were used in identifying the heat waves by the authors of [6] cited above.

In this study, the heat (cold) wave is considered to mean a change in average daily air temperature by three degrees or more during at least two days. Note that if against the background of the heat/cold wave a cooling (warming) or isothermy lasted for one day, such waves were regarded as one heat (cold) wave.

Source material for identifying waves was represented by data on average daily air temperatures posted on the website operated by VNIIGMI-WDC (http://meteo.ru/). Air pressure fields at sea level were constructed by using NCER/NCAR reanalysis data. The workspace was represented by the territory liited by the meridians 40°–180° E and 40°–70° N. Synoptic analysis used weather maps for the period 00 h GMT.

The operation algorithm of the model is as follows:

1. Preparation of the working sample.
2. Determination of a typical (reference) field.
3. Plotting predictive dependencies.
4. Validation of the model operation.

Archival samples of pressure fields are generated for each particular month separately for the cases of heat and cold waves. It should be remarked that two fields are selected for each wave: the day when the extremum of the wave occurs, and the day preceding the extremum of the wave. In accordance with this, in each month we obtain a separate sample for the cases of an abrupt warming/heat wave, and a separate sample for the cases of an abrupt cooling/cold wave. An abrupt warming/cooling is taken to mean a rise/decrease in average daily air temperature by 8° or more lasting for two days or more.

The second stage involves identifying the reference field which is used as the predictor afterwards. By determining the reference field, we solve the problem of selecting the most typical synoptic situation where there occur abrupt differences in average daily air temperature. The reference field is determined by the Euclidean distance formula (formula 1). The field that is found to be closest, according to the Euclidean distance, to all the other fields of the archival sample is considered to be a reference field:

$$C(i,p) = \frac{1}{q}\sqrt{\sum_{j=1}^{q}(x(i,j) - x(p,j))^2}, \tag{1}$$

where $C(i,p)$ is the proximity (remoteness) parameter between the ith and p-th fields; $x(i,j)$, $x(p,j)$ are the values of air pressure at sea level at the jth point of the ith and pth fields; k is the number of fields being compared, and q is the number of nodes on a regular grid.

Formula (2)

$$S(i) = \frac{1}{k}\sum_{p=1}^{k} C(j,p), \tag{2}$$

is used to estimate the similarity of the ith field with all the other fields of the working sample. From the least value of the parameter $S(i)$ it is concluded which of the ith fields is closest by distance to all the p-fields.

Plotting of the predictive dependence implies determining the analog field to each reference field. Selection is done by the principle of "floating analog" allowing for a displacement of the analog fields relative to the reference fields: by ±15° along the meridian (Δφ), and by ±60° along the circle of latitude (Δλ). Note that selection of the analog uses a less strict (compared to the proximity/remoteness parameter *C(i,p)*) indicator of similarity ρ [13, 14].

The bimonthly interval as substantiated by V.F. Martazinova [9–12] has become an optimal time interval for selection of analog fields. Note that the quasi-bimonthly rhythmicity of atmospheric processes is well known. Its existence is confirmed in many scientific publications, such as [25].

A forecast is made in accordance with the resulting analog fields. To accomplish this, it is necessary to estimate the degree of proximity/remoteness of the analog fields with daily baric fields of the month separated by one month backward from the predicted one. From the least values of the proximity/remoteness parameter *C (i, p)* we determine the dates of extremums of heat and cold waves according to the bimonthly quasi-periodicity of atmospheric processes. Thus the forecast lead time is found to equal one month.

3 Results and Discussion

Samples of baric fields forming abrupt warmings and abrupt coolings were generated for Irkutsk for January. In each sample, formulas 1 and 2 were used to calculate reference pressure fields characterizing the cases of warmings and coolings. Reference fields were determined by using data for the period 2010–2015: a training sample. The control sample included the years 2016 to 2021. Table 1 presents dates of the January reference fields.

Table 1. Reference pressure fields of heat and cold waves

Month	Date of reference field	
	Heat waves	Cold waves
January	January 28, 29, 2010	January 19, 20, 2012 January 27, 28, 2015

According to the calculations presented in Table 1, the coolings in the winter (January) are determined by two types of synoptic processes. A strong cooling occurs when the observation point (Irkutsk) is located in the central part of the winter Asian anticyclone. The reference field of such a process is presented in Fig. 1a, showing that Irkutsk is under the influence of the central part of the anticyclone. A cooling that is supported by such a synoptic situation lasted in Irkutsk for six days, and the temperature decreased from –5 °C to –25 °C.

Based on calculations of the reference fields (Table 1), it was found that cold waves in the winter in Southern Siberia (Irkutsk) are also formed during inflows of the cold air along the eastern periphery of the Asian anticyclone (type II). The reference field of

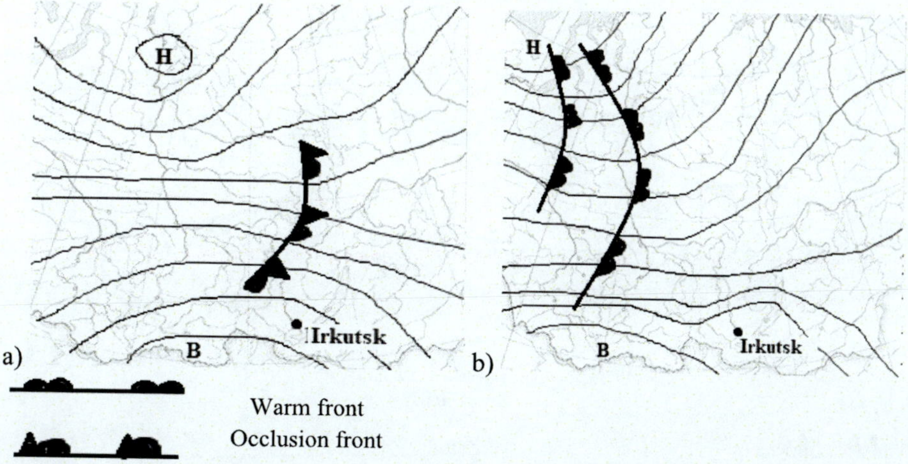

Fig. 1. Fragments of the surface map a) January 27, 2015, b) January 28, 2015. Black lines show isobars, letters H – cyclone, and B – anticyclone

such a process is given in Fig. 1b. In this case, the cold wave was shorter in duration than the preceding wave, and the average daily air temperature decreased from –12 °C to –27 °C.

Relativ warmings in the winter in Southern Siberia also occur quite often. On average, in the case of heat waves the temperature rises from –20 °C to –25 °C to –10 °C … -5 °C. As the analysis of the calculated dates of r3eference fields showed, the winter warmings in Southern Siberia are determined by the arrival of relatively warm air masses from the west across the southern portions of extensive depressions, with the adjacent northern periphery of the anticyclone located to the south of it. An example of the reference field of an intense heat wave is presented in Fig. 2a, b. As is evident from the figures, an intense warming was evolving with the arrival of warm air masses from the west along the periphery of the winter Asian anticyclone, with a northern powerful cyclonic depression originating on the Arctic front. With such a type of process, a warming occurs quite abruptly. In the present case the temperature increased from –18 °C to –7 °C during two days.

Thus the synoptic analysis confirms the correctness of identification of reference fields by the method used in this study. The correctness of the reference fields as a result of independent tests using data for the period 2016–2021 (independent sample) showed 81% for heat waves, and 79% for cold waves.

For each month we made forecasts of abrupt warmings and abrupt coolings using independent sample (2010–2015) and an independent sample (2016–2021). In the stage of verification, the heat/cold wave was considered correct subject to the following conditions: 1) the difference in average daily air temperature should be five degrees or more, and 2) the time of occurrence of an extremum of the wave either should coincide with the day of forecast or the shift should not exceed two days. As an illustration we provide a forecast of abrupt changes in average daily air temperature as of January 2017 (Table 2).

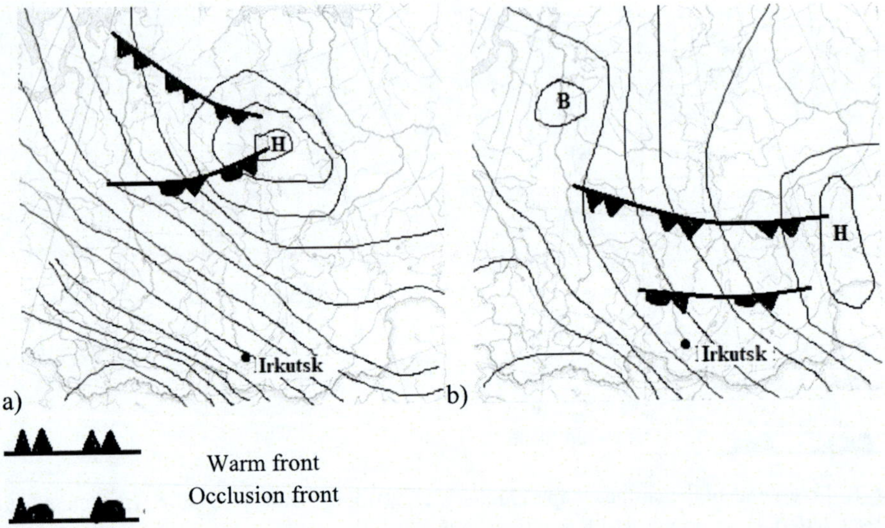

Fig. 2. Fragments of the surface map. a) January 28, 2010, b) January 29, 2010. Black lines show isobars, letters H – cyclone, and B – anticyclone.

Table 2. Forecast of abrupt changes in average daily air temperature (January 2017).

Heat waves			Cold waves		
Forecast	Fact	Correctness	Forecast	Fact	Correctness
–	2–3	–	8–9	9–19	+
13–14	14–15	+	17–19	19, 21	+
24–25	30–31	+	28–29	27–28	+

According to Table 2, two heat waves and three cold waves were forecasted for January 2017. In actual fact, January 2017 showed three heat waves and three cold waves (Fig. 3). Thus only one heat wave among all the waves was not forecasted. Perhaps this is due to the fact that t4he wave began to evolve in the preceding month, because the warming maximum occurred on January 2–3. Note that the waves whose extremums occur in the early days of the month are forecasted somewhat worse than over the course of the remaining part of the month [13, 15, 17].

Interestingly enough, the cold wave occurring in the middle of the month was successfully forecasted, in spite of the fact that this cold wave was interrupted by a one-day rise of temperature. Note that a coldwave is very frequently accompanied by short-lasting one-day increases in temperature, usually associated with the cyclogenesis processes – generation of the wave on the cold front [13, 15, 17].

Calculations showed that the correctness of forecasts by using an independent sample made up 73% for abrupt warmings and 71% for abrupt coolings. In the case of independent tests, the correctness of forecasts of abrupt warmings and abrupt coolings

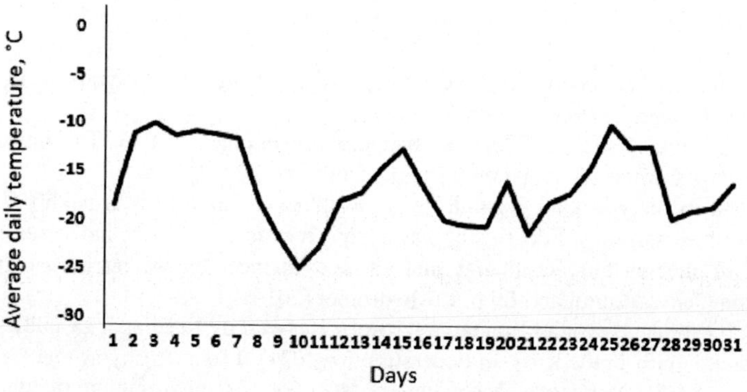

Fig. 3. Variation in average daily air temperature in January 2017.

made up 71 and 69%, respectively. Based on the foregoing and taking into consideration a fairly large forecast lead time – one month, it may be assumed that the suggested technique of forecasting abrupt differences in average daily air temperature can be used quite successfully in solving similar forecasting problems in different regions.

4 Conclusions

The research presented in this paper suggests the following conclusions:

1. A validation of the operation of the long-term forecasting model for extremums of meteorological parameters has been carried out by considering an example of forecasts of average daily air temperature for Southern Siberia in January.
2. The study revealed the most characteristic, reference synoptic processes generating abrupt warmings and abrupt coolings in Southern Siberia in the wintertime by using the Euclidean distance formula. The validity of using precisely this formula in determining reference fields was confirmed by synoptic analysis.
3. Cold waves in the wintertime are determined largely by two types of processes. One type implies that strong coolings are associated with a radiation cooling near the center of the winter Asian anticyclone; the other type implies that strong coolings are generated by advection of air masses along the cold eastern periphery of the anticyclone. Heat waves in the winter are associated with advection of warm air masses across the southern portion of an extensive depression adjacent to the northern periphery of the anticyclone located further to the south.
4. The correctness of forecasts using an independent sample made up 71 and 69% for heat and cold waves, respectively, which can be considered quite successful in view of the rather stringent criterion of forecast correctness of waves and the one-month forecast lead time.

References

1. Intergovernmental Panel on Climate Change, Climate change 2014. Synthesis Report, 151 p. Cambridge University Press, Cambridge (2015)
2. Intergovernmental Panel on Climate Change, Climate change 2013. The physical science basis. 3676 p. Cambridge University Press, Cambridge (2021)
3. Vtoroy otsenochnyy doklad Rosgidrometa ob izmeneniyakh klimata i ikh posledstviyakh na territorii Rossiyskoy Federatsii. Obshcheye rezyume (The second assessment report of Roshydromet on climate change and its consequences on the territory of the Russian Federation.General summary, 60 p. Roshydromet (2016)
4. Doklad ob osobennostyakh klimata na territorii Rossiyskoy Federatsii za 2021 god (Report on climate features in the Russian Federation for 2021), 110 p. Roshydromet (2022)
5. Alimpieva, M.A., Morozova, S.V.: On the state of the environment on dangerous climatic trends. In: International Scientific and Practical Conference, pp. 383–387. (In Russ). Nauchnaya kniga, Voronezh (2017)
6. Kruglova, E.N., Kulikova, I.A., Tishchenko, V.A., Khan, V.M.: Prediction of heat waves on intra-seasonal time scales, no. 348, pp. 95–105. (In Russ). Tr. Hydrometeorol. Nauchn. Tsentra (2019)
7. Pepler, A.S., Diaz, L.B., Prodhomme, C., Doblas-Reyes, F.J., Kumar, A.: The ability of a multy-model seasonal forecasting ensemble to forecast the frequency of warm, cold and wet extremes. Weather Climate Extrem. Release **9**, 68–77 (2005)
8. Barnston, A.G., Mason, S.J.: Evaluation of IRI's seasonal climate forecast for the extreme 15% tails. Weather Forecast. Release **26**, p. 545–554 (2011)
9. Martazinova, V.F.: Changes of atmospheric circulation in central-east Europe at recent decade and long-range forecasting. In: Proceedings of the Workshop on Regional Climate Change in Central-East Europe, pp. 1–6 (1994)
10. Martazinova, V.F., Mladinaya, L.V.: The method of detailed weather forecast for a month according to the method «floating analogue», release 219. pp. 46–50. (In Russ.). Tr. UkrNIGMI (1986)
11. Martazinova, V.F., Ivanova, E.K.: Method for forecasting air temperature and humidity in Ukraine for natural synoptic periods on the territory of Ukraine, release 233, pp. 83–87. (In Russ.). Tr. UkrNIGMI (1989)
12. Martazinova, V.F., Sologub, T.A., Ivanova, E.K., Matsak, V.V.: Anomalous atmospheric processes over Ukraine and their forecast with a monthly lead time, release 239, pp. 63–68. (In Russ.). Tr. UkrNIGMI (1991)
13. Morozova, S.V.: Forecast of heat and cold waves for the Saratov region using the physical-statistical method of V.F. Martazinova «floating analogue», release 363, pp. 138–159. (In Russ.). Gidrometeorologicheskiye prognozy (2017)
14. Morozova, S.V.: Technology of long-term forecasting of precipitation periods for the Saratov region, no. 35, pp. 145–158. (In Russ.). Trudy Gidrometeorologicheskogo nauchno-issledovatel'skogo tsentra Rossiyskoy Federatsii (2015)
15. Morozova, S.V., Polyanskaya, E.A., Kononova, N.K.: On the possibility of long-term forecasting of abrupt changes in weather conditions on the example of Voronezh. In: International Scientific and Practical Conference, pp. 337–342. (In Russ). Nauchnaya kniga, Voronezh (2021)
16. Duytseva, M.A., Ped, D.A.: Peculiarities of cold and heat waves on the European territory of the USSR, release 123, pp. 34–62. (In Russ.). Tr. CIP (1963)
17. Morozova, S.V., Alimpieva, M.A., Zavyalova, E.V.: Sharp changes in the average daily air temperature and the possibility of their forecast. In: II International Scientific Conference in Memory of Corresponding Member of the Russian Academy of Sciences D.G. Matishova, pp. 58–62. (In Russ). Rostov-on-Don (2020)

18. Perkins, S.E., Alkexander, L.V.: On the measurement of heat waves. J. Climate Release **26**(13), 4500–4517 (2013)
19. Robinson, P.J.: Definition of a heat wave. J. Appl. Meteorol. Climatol. Release **40**, 762–775 (2000)
20. Sreznevsky, B.I.: Cold waves from Novaya Zemlya to Persia, India and the Irkutsk region. Meteorologicheskiy vestnik **4**, 107–110 (1899)
21. Adrianov, L.V.: Features of the distribution of cold and heat waves in extremely warm and extremely cold Junes in Saratov. Voprosy klimata i pogody Nizhnego Povolzh'ya, release **7**, 78–82 (1971)
22. Arkhangelsky, V.L., Kotova, L.M.: Waves of cold and warmth in the troposphere and stratosphere over the Lower Volga Voprosy klimata i pogody Nizhnego Povolzh'ya, release 4, pp. 35–45 (1966)
23. Khrabrov, Yu.B.: Study of the calendar features of cold weather and sweatplace in Moscow. Trudy TSIP, release 150. 158 p. (1967)
24. Gruza, G.V., Ran'kova, E.Ya.: Assessment of the possible contribution of global warming to the generation of extremely hot summer seasons in the European territory of the Russian Federation. Izvestiya RAN. FAO, vol. 47, no 6, pp. 717–721 (2011)
25. Vil'fand, R.M.: Classification of air pressure fields on the territory of the 1st natural synoptic region. Trudy Gidromettsentra, release **274**, 3–11 (1986)

On the History and Prospects of the Westernist Idea in Russia

Pyotr L. Popov(✉) and Alexey A. Cherenev

V.B. Sochava Institute of Geography, Siberian Branch, Russian Academy of Sciences, Ul. Ulan-Batorskaya 1, 664033 Irkutsk, Russia
plp@irigs.irk.ru

Abstract. This article addresses the issues of cultural-historical identify of Russia and, specifically, Westerism, one of the geopolitical concepts. Accordwing to it, Russia is a part of the European civilization. The significance of these issues in the history of Russian political thought is traditionally great, which is due to the special features of the geographical location of the country, and to its common features with countries of Eastern and Western Europe and Central Asia. An analysis is made of the historical aspect of the development of the ideas of Westernism and its relation to other concepts of the civilizatonal position of Russia. The borrowing of Western technologies entailed assimilation of many elements of the lifestyle and influenced the world outlook sphere. The emergence of the concept of Slavophilism became the response to the negative sides of these changes. Over the course of its hstorical development and infeluence of the political realities upon it, the Westernism separated into the conservative, left revolutionary and liberal direcions. In the latter half of the 20th century, the liberal direction acquired colonial features. In the 19th century there emerged another concept alternative to the Westernism: Eurasianism; in the late 20th century, the isolationism shaped itself conceptually. At present, the world outlook of the political elites of Russia is characterized by a combination of the elements of the Westernism and isolationism.

Keywords: Westernism · Slavophilism · Eurasianism · Isolationism · Cultural-historical identity

1 Introduction

All or almost all sqociological concepts have a clearly pronounced geographical aspect. Identification of factors influencing the socium, as claimed by the sociological cncept, calls for inter-country and inter-regional comparisons. Therefore, in what follows when dealing with the socio-geographical concepts, we also mean both the actually socio-geographical and partly sociological concepts. Sociological concepts are of two types: globally oriented, and regionally oriented. The globally oriented concept claims to reflect the regularities occurring globally. A reguarity implies the relationship between the factor and the state of he socium. The global concept seeks to reveal the main factors (global universals) determining the main characteristics (which are also global universals) of

I. N. Vladimirov et al. (Eds.): RERSDNA 2022, SPEES, pp. 142–150, 2023.
https://doi.org/10.1007/978-3-031-28978-1_16

the state of the socium. The regionally oriented concept claims to reflect the regularities operating at the scale of a region of one or another level and, as a rule, is a regional application of the globally oriented concept. The regionally oriented sociological concept is directly adjacent to social geography. The problematics of determining a cultural-historical identity of its own, and the position in the world system emerged in many countries; but, perhaps, for social thought in or country it has been and is particularly meaningful. Russia is geographically close to several cultural-historical communities, or regions of the world (at this stage we do not go to the heart of the differences of these related notions). Namely, to Eastern Europe, especialy to Slavic and Orthodox countries; to a lesser extent, to Western Europe, partly via countries of Northern Europe; to non-European eastern-Christian countries; to countries of Central Asia, largely Muslim countries, and to countries of Easst Asia (China, Korea and Japan). As a large and sufficiently independent geopolitical actor, Russia entered into diverse, including conflict, relationships with its neighbors, which strengthened attention to the theme of cultural kinship, because geopolitical communities, unions, cultural-historical communities and civilizations are interrelated, though not identical realities. Special attention was paid to the teme of knship (or its absence) with the West, because it is from the West the main cultural influences on our country have been coming over the last 500 years and it is the West where the main military adversaries and allies resided. Historically, the civilizational problematics became more sophisticated: at the beginnng, it was limited to disputes between Westerners and Slavophiles (supporters of the two concepts oriented toward Europe) and, in connection with the growth of the significance of Asia, there emerged the Euroasianism and, finally, in modern Russia, the isolationist motives shaped themselves conceptually, which were always present in Russian social thought. In modern Russia there remain the traditions of all the aforementioned concepts.

Modern Russia, within today's boundaries, in today's social system and ideological situation, is a historically new status of our country, and this makes the problematics of the civilizational self-determination still more relevant. In this paper, we try to characterize the history and prospects of one of the direction in this circle of ideas, the Westernism.

2 Material and Methods

The source material was provided by publications of leading Russian scientists, geographers, historians and political scientists – both contemporary authors and classics whose work still remains relevant. The information base for interpretation of historical events and their consequences included publications of V.O. Klyuchevskii, A.I. Geotsen, and A.P. Bogdanov. We also examined publication of contemporary authors on globalistics and civilizational problematics: R. Robertson, M. Waters, M.F. Guillén, R. Kiely, P. Marfleet, Z. Bauman, P.N. Savitskii, N.S. Trubetskoi, V.L. Tsimburskii, and V.A. Nikonov. We also took into consideration data on the origin of the population of Russia, based on DNA genealogy. The research was based on analyzing the concepts of the civilizational position of Russia and was the continuation of our work on the Eurasianism and isolationism.

The statistical base for this investigation was provided by data in the report on world development for 2021.

This research is based on using the comparative-geographical and historical methods and the civilizational approach. Use was also made of the methodological approaches of political-geographical generalization and analysis.

The concepts were analyzed with due regard for the following principles of their construction.

In constructing a concept, the factors are usually subdivided into two groups: little significant and highly significant. In this (in such a crude dichotomy) lies the difficulty of a quantitative assessment of the elements of the social reality. Actually, in constructing the concept there occurs the accentuation of one group of factors and the neglect of the other. Accordingly, the accentuation of a certain group of factors or phenomena (the indication of which confirms the action of the selecsted factors) and neglect of the other group of facts or phenomena (the indication of which is neutral relative to the concept or contradicts it).

Accentuation implies the use of the means with the purpose of increasing the significance of a certain group of facts in the conscience of the addressee of the concept. Such means imply: 1) the indication of these facts (facts that confirm the concept), and и 2) advancing the thought, according to which the facts of such a type are especially significant in a given context; it is the application of the sociological concept (as a rule, globally oriented) to the socio-geographical concept. For instance, Marxism (a globally oriented sociological concept) can be used in the creation of the Marxist concept of the history of Russia..

Neglect imies the use of the means with the purpose of decreasing the significance of a certain group of facts in the conscience of the addressee of the concept. Neglect is also realized 1) simply by the non-indication of these facts – it is the disregard, or 2) advancing the thought, according to which the facts of such a type are of little significance in a given context.

In this paper, we have analyzed the facts accentuated in different concepts of the Westernism.

3 Results and Discussion

In the new history of Russia, the question of the attitude to the West as a cultural whole most likely arose for the first time in the 16th–17th centuries. Note that before the reforms of Peter the Great but in the process of their historical maturing. At that time it became clear how much the West took the lead in the military-technological sphere with respect to Russia (Rus) and with respect to the rest of the world. "The Government began to ask forifoeiers for assitance primarily in order to satisfy their most pressing material needs concerning the defense of the country and military affairs – it is in these areas that the backwardness was particularly critical. It borrowedfrom other countries military and, later, other technical improvements reluctantly, without looking far ahead to see the possible consequences of such endevours and without looking into the scope of efforts made by the West-European intellect to reach such technical successes and what view of the tasks of world outlook and what tasks of being governed those efforts" [1, p. 485]. The influence was gradually strengthening to encroach upon the other spheres, including – and it was particularly sensitive – the world outlook. The most conservative circles of Moscovite Rus saw in this a threat to the religious traditions. An expression of protest against the western influences was the church sqchism of the 17th century. A direct cause for he sshism were the ritual reforms in the Russian Orthodox Church but the schismatics (Old Ritualists) saw in them a sign of a whole system of innovations. Old Ritualism might have something in common with Slavophilism of the 19th century; but a difference deserves attention. Characteristic for the Old Ritualism was not the common-Slavic and even not Russian (East Slavic according to current terminology) but Great Russian self-consciousness.

Russia was turning from a state simiar to the type of national state into an empire. The guardians gravitated toward peacemaking, and the reformers, "progressives" toward the imperial mentality [2]. The formation of the link of the Westernism with military activity, with the idea of an emp[ire was promoted by several factors: techological innovations coming from the West were intended primarily for the military sphere and provided Rus with an advantage over its eastern adversaries, thuscreating preconditions for the territorial expansion; Little Russia and Belorussia gravitated toward a unification with Russia, and the realization of this plan which was favored by the westernist orientation of a part of the Moscow elite (these countries tied, to some extent, Moscovite Rus with the rest of Europe) led to a collision with Poland; at that time the countries of the West began to transform to empires. A new quality was acquired by the western influence at the epoch of the reforms of Peter the Great when Moscovie Rus turned into the Russian Empire. The development of Russia along this path in t4he 18th–19th centuries involved considerable achievements and a number of losses and negative phenomena. The defensive burden on the country increased; the position of landlord's peasants worsened; a profound cultural split emerged between the portion of the socium subjected to the western influence (primarily the nobility) and its other part – such are some of the negative phenomena. The reaction involved the direction of thought that was given the name Slavophilism. In the dispute with him, the westernist cultural orientation was finally shaped conceptually in the 19th century, and the term "westernism" itself was coined. V.O. Klyuchevskii expressed the "essence of the two views in this last stage of their development thus. The Westernists taught: according to the bases of our culture, we are Europeans, only younger in our historical age; therefore, we must follow the path traveled by our older cultural brothers, assimilating the fruits of their civilization. Yes, the Slavophiles objected, we are Europeans, but eastern ones, we have our vernacular origins of life, and we are to develop them with our own efforts without being tethered to Western Europe. Russia is not a disciple and not a companion of Western Europe, it is its successor" [1, p. 484]. The Westernism ignores (considers little significant) such phenomena as kinship of Russians with other Slavic peoples, and the specific character of Slavic culture with respect to West-European culture.

At the epoch under consideration the ruling circles of Russia ideologically gravitated toward the support of the conservative forces of Europe against the revolutionary (liberal and socialist) forces. This circumstance did not prevent the collision of Russia with the conservative Germany and Austria-Hungary and the union with the more liberal Great Britain and France during the First World War. This is one of the examples illustrating the divergence of the ideological and geopolitical relations.

In the interstate conflicts of the 18th–19th centuries, Russia most often countered not Europe as a single entity but one or another part of Europe in alliance with another part; the allied part of Europe, as a rule, was larger than the hostile part. For instance, during the Napoleonic Wars France fought agaist most of foreign Europe and, although it subjugated Europe for some time), the new European empire was fragil and collapsed under the influence of the defeat of the French Army in Russia, the involuntary allies of France turned their weapons against it. It is pertinent to make mention of this well-known circumstance, because at present Napoleon's invasion of Russia is often treated as the collision of Russia with the collective West. The sole exception is provided by the Crimean War when Russia found itself in isolation.

The intention not only to catch up but also overtake the West was not foreign not only to the Slavophilism but also the Westernism, especially the left one. A.I. Herzen [3] pointed out the incapaciy of Europe for the social upheaval, bearing a revolution of socialist nature in mind (which would be a new step of progressive development); like the other Russian leftists, he believed in the possibility of such a revolution in Russia. The Marxists who gained a predominance in the Russian socialist movement firmly gravitated toward the Westernism. Classical Marxism, though it is a class-globalist concept (Socialism must triumph in all countries of the world), has the eurocentric aspect, because the victory of Socialism was thought to occur initially in the most developed countries of the West; it is easy to conclude that not only will these countries strengthen with the advanced system their leadership, but they also will lead (not only by force of example) the rest of the world. Marxism is the Eurocentric theory also in the sense that the facts on which it relies, including an important section, such as the characteristic of the socio-economic formations, were taken largely from the realities of the European civilization; interpretation of some non-European realities in Marxism is diccult (for instance, the "Asian method of production"). K. Marx and F. Engels saw in Russia primarily the source of a threat to the European revolution. However, once the socialist revolution in Russia succeeded and no continuation of it followed in the West and the Soviet Union tirned out an only country building Socialism, there emerged preconditions for an isolationist tendency in the Soviet ideology.

After the October Revolution our coutry, the former stronghold of the conservative reaction, turned into a state supporting the Socialist movement abroad, including (and especially in the beginning) in the West. The influence of the Socialist movement in the West and, still more, the contradictions btweenthe leading countries of the West, were weakening the foreign-policy separateness of the Soviet Union. After the Second World War, the bloc under the control of the Soviet Union contained most of the countries of Eastern Europe, including Slavic countries; this promoted a strengthening of the westernist basis in the world outlook of the Soviet elites. On the other hand, Eurasianism also acquired a great influence. This concept emerged in the 1920s in the milieu of white emigration. According to the classics of the Eurasianism [4–6], Eastern Slavs are civilizationally closer to the peoples of Central Asia than to foreign Europeans, including Western and Suthern Slavs. These similarities and differences are derived from the physical-geographical location (in accordance with the ideas of western geopolicy, the opposition between the "Sea" implying the "West" and Land, i. e. "Eurasia", is proclaimed. The Eurasianism (in spite of the western roots) is fundamentally farther from the Westernism than the Slavophilism (Russia is not a continuer of Europe, because it shaped itself in a different "development place").

In the Slavophilism, the religious component is strong; at the epoch of pedominant atheism this factorв weakened the influence of its ideas. The Eurasianism, however, in the materialist spirit, accentuates the culture-forming significance of the natural environment. However, the ruling circles of the Soviet Union were clearly dominated by the westernist vector, in accordance with the tradit4ions of the country and the theory of Marxism.

The weakening and disintegration of the Soviet Union led to the emergence (or a radical srengthening) of the liberal offshoot of the Westernism in which there appeared features differring from the former variants of the Russian Westernism. Emphasis was placed not on the closeness of Russia to the West but on the differences of Russia from the West (disadvantageous for Russia). The origrinaion of such differences is attributed to the influence of Asia and the unfavorable natural environment. More likely the specific character of Russia is treated not as a provisional backwardness but as a wrong development (which would require a more radical uphaval to the Russian traditions and mentality). The historical insolvency of Russia combines with the physical-geographical inferiority ("Eurasian wasteland"). The liberal Westernism gravitates toward the idea of redundancy of the Russian space and, accodingly, toward the need for its "drawdown", a reduction (and at the expense of the Asian part), and toward the idea of refeormingr the administrative-territorial division with the purpose of transforming macroregions andnational republics of Russia to relatively independent (economically and politically) entities. The internal irregularities of Russia are emphasized, whereas the West, on the contrary, is perceived and a monolith. The differences of Russia from the other Slavic countries are highlighted, their specific character with resqpect to Western Europe is ignored as well as ignoring the anti-Slavic (and not only anti-Russian) traditions in West-European thought.

The liberal-colonial Westernism was influential in the Russian publicistic, educational and scientific literature, and in mass media of the 1990s. The historical background favoring the Russian Westernism in its liberal form was represented to a unipolar world and the associated flourishing of the ideas of globalism [7–11]. In the 2000s, the political elite of Russia, becoming convinced that no decent integration with the West was possible and not accepting a neocolonial dependence, moved to a more independent external and internal policy. This was also manifest in concepts interpreting the position of our country in the system of world civilizations.

By the 2000s, in the international situation of Russia there had occurred significant changes which inevitably affected conceptual activity. The relationships with the West have deteriorated. There occurred a change in the image of the West in Russia. Almost al Slavic countries and almost all Orthodox countries found themselves involved into the orbit of the West. Although the Central Asian countries preserved partnership with Russia within the framework of post-Soviet entities but were also moving closer Turkey belonging to the Islamic world and to the Western NATO alliance. Unipolarity of the world has become not obvious, and the tendencies for deglobalization have ben enhanced [12]. These circumstances have not favored either Westernism (especially in its liberal-colonial variant) or Slavophilism or Eurasianism. The positions of isolationism have strengthenes, in acocrdance with which Russia is a special civilization [13]. At present this position is often proclaimed also by political feagures and political scientists [14].

At the same time, there also remain some tendencies promoting weesternism. A rapid growth of the non-European centers of power, and a massive penetration of migrants from non-European countries give cause for concern both in the West and in Russia, especially since in the immediate vicinity to it there are some of such centers (China, Japan, and Turkey) [15]. This circumstance provides a historically profound basis for the approchement of Russia and the West. According to the set of signs determining culture (reigion, language, composition of the gene pool, science and technologies, legal system, arts, and natural environment), Russia is closer to European countries, primarily Slavic ones, than to Asian countries. This would bring up the questions of whether our country should be assigned to Europe on the rights of some of its peculiar portion or should be considered a special civilization or should the Slavic or Orthodox Slavic countries (including Russia) in the European civilization or should they be considered a separate civilization, but it is quite clear that the Islamic world, China, Japan, Southeast Asia, Latin America and Africa are farther from Russia than Europe. Recent discoveries in genetics that gave rise to DNA genealogy confirmed the weakness of Asian influences in Russia, the kinship of most of its population largely with Slavic as well as other European peoples. Moreover, the development of this scientific area also strengthened the positions of the hypothesis about massive migration from the territory of modern Russia to the west in the Bronze Age; the motives of kinship and influence of the West on Russia incorporate the motive of the historically profound influence of ancestral Russia on Europe [16].

A certain combination of the elements of isolationism and westernism can be noted in the position of the political leaders of Russia, and in the expert community. V.V. Putin has repeatedly spoken about Russia as a country-civilization [17], but he also emphasized its connection with Europe. "We hoped that the end of the Cold War would be a comon victory for Europe. It seemed a little more, and that Charles de Gaulle's dream would become a reality, namely about a united continent, even not geographical "from the Atlantic to Ural" but about cultural, civilizational – from Lisbon to Vladivostok". "...all post-war history of the Greater Europe confirms: the prosperity and security of our common continent are only possible through combined efforts of all countries, including Russa. Because Russia is one of the major European states. And we feel our inseparable cutural and historical connection with Europe" [18].

In modern Russia, the image of our country as a stronghold of traditional values is taking an increasingly greater place. The values are also traditional for the West from which it is moving away increasingly. These values and traditions were, to a great extent, created by Christendom: the family in the traditional sense, certain assessments of cultural and political phenomena of the past (with all their diversity and, sometimes, incompatibility, they have quite a few basic, common elements), accentuation of the logos and verballogic thinking as compared to the intuitive-subconscious sphere.

With such an understanding, our country is perceived as an important part of the European civilization defined, however, within broad geographical bfoundaries.

4 Conclusions

In the modern world, apart from the aforementioned growth of the non-European centers of power, there is also a tendency toward the violation of the unity of the West, which manifests itself both at the interstate and intrastate levels (Brexit). A likely enhancement of contradictions between traditionalists and globalists in the West may entail a strengthening of the traditional westernist tendency in Russia and a weakening of the liberal-westernist circle of ideas. A critical strengthening of one of the non-European centers of power may lead to the rapprochement of Russia and t4he West, to a new enhancement of westernism in Russia, and to a weakening of contradictions between traditionalists and globalists in the West and, accordingly, to the emergence of a new version of westernism in Russia.

But these prospects are not near-term ones. The Asian centers of power, still in many respects, are far off the level of development of major Western countries. The economic weight of China and India is, in many respects, determined by the size ofe their population. As regards the level of development of human potential, the determination of which takes into account the level of economic development, education and duration of life, China and India occupy the 85th and 131st places, respectively (as of 2020). Such a backwardness hampers the putting of the foreign-policy weight of a country into correspondence with the scale of its economy.

One can expect a long-term continuation of the current international situation in its main particularities, and a long-lasting preservation of a combination of westernist and isolationist features in Russian thought. Culturologically, more likely скорее westernist features, and geopolitically, more likely isolationist features.

Acknowledgtments. This research was done under scientific projects AAAA-A21–121012190018-2 and AAAA-A21–121012190056-4.

References

1. Klyuchevsky, V.O.: Russian History. Eksmo, Moscow (2020)
2. Bogdanov, A.P.: From Chronicle to Research. RISK, Moscow (1995)
3. Herzen, A.I.: Past and Thoughts. Khudozhestvennaya literatura, Moscow (1969)
4. Savitsky, P.N.: Continent Eurasia. Agraf, Moscow (1997)
5. Trubetskoy, N.S.: About the Turanian element and russian culture. Russia between Europe and Asia: Eurasian temptation, pp. 59–77 (1993)
6. Trubetskoy, N.S.: Pan-Eurasian nationalism. Russia between Europe and Asia: Eurasian temptation, pp. 90–100 (1993)
7. Robertson, R.: Globalization: Social Theory and Global Culture. SAGE Publications, London (1992)
8. Waters, M.: Globalization. 2nd ed. Routledge, London (2001)
9. Guillén, M.F.: Is globalization civilizing, destructive or feeble? a critique of five key debates in the social science literature. Annual Review of Sociology **27**(1), 235260 (2001)
10. Kiely, R., Marfleet, P.: Globalisation and the Third World. Routledge, London (1998)
11. Bauman, Z.: Globalization: Human Consequences. Columbia Univ Press, New York (1998)

12. Cherenev, A.A., Popov, P.L.: Asian Russia and the deglobalization trend. Earth and Environmental Science: IOP Conference Series **629**, 012084 (2021)
13. Tsymbursky, V.L.: Island Russia. Geopolitical and chronopolitical works. Rosspen, Moscow (2007)
14. Popov, P.L., Cherenev, A.A., Saraev V.G.: On the concept of civilizational isolation of Russia. Izvestiya IGU Seriya «Politologiya. Religiovedeniye» **33**, 17–24 (2020)
15. Popov, P.L., Cherenev, A.A.: To the problem of objective prerequisites for Siberian separatism. Bulletin of the Tomsk State University. Story **77**, 51–60 (2022)
16. Belanovskaya, E.V., Belanovsky, O.P.: Genetic traces of historical and prehistoric migrations: continents, regions, peoples. Vestnik Vavilov Society of Geneticists and Breeders **13**, 401–408 (2009)
17. Putin, V.V.: The future of Russian civilization depends on success in the development of high technology. https://nauka.tass.ru/nauka/8493647?utm_source=yxnews&utm_medium=mobile/amp. Accessed 1 June 2020
18. Putin, V.V.: Sei offen, trotz der Vergangenheit. Die Zeit (2016)

Check for updates

Industrial Production Potential and Foreign Economic Relations of Eastern Siberia

Vera A. Rudneva(✉)

Irkutsk Scientific Center SB RAS, Lermontov Str., 134, Irkutsk 664033, Russia
varud8892@yandex.ru

Abstract. The author calculated the production potential for the regions of Eastern Siberia between 2010 and 2018. Three groups of indices reflecting the level of industrial development in the territories under consideration were identified on this basis. It has been found that the potential decreases mainly from west to east, and its fluctuations reflect investment activity in the macroregion over the period of observations. The exception is Krasnoyarsk Krai, which is strengthening its position in all respects. It has the greatest potential for industrial diversification. The remaining subjects are characterized by the volatility of their values in relation to the leader. Of particular importance is the indicator characterizing manufacturing industries. In Buryatia, Krasnoyarsk Krai, and Irkutsk Oblast, there is a high proportion of high-added value products due to mechanical engineering. The assessment of foreign trade relations of the subjects under study was carried out taking into account the regional economic indicators. It was revealed that the internal regions are more successful in foreign economic activity than the border ones. An analysis of the markets showed that Krasnoyarsk Krai, Irkutsk Oblast and the Republic of Khakassia are focused on the export of low-added value products, while the border regions supply goods from industries based on the exploitation of natural resource rent. Mostly Asian countries purchase raw materials and products of primary processing. The import structure is similar to the all-Russian one and is aimed at the purchase of high-tech goods. In Eastern Siberia, there is a large proportion of founders from China, not only in industry, but also in other sectors (especially in Zabaikalskii Krai). The eastern vector of Russia's economic development makes the area under study one of the key actors in strengthening foreign non-economic activities with the Asia-Pacific countries.

Keywords: Production potential · Industrial development · Branches of specialization · Processing of raw materials · Foreign economic relations · Export markets · Region

1 Introduction

This article considers Eastern Siberia within the framework of the Angaro-Yenisei macroregion (Krasnoyarsk Krai, Irkutsk Oblast, Republics of Khakassia and Tuva) and Transbaikalia (Republics of Buryatia and Zabaikalskii Krai). The latter is due to long-standing economic ties that were activated during the period when the units were part of

I. N. Vladimirov et al. (Eds.): RERSDNA 2022, SPEES, pp. 151–159, 2023.
https://doi.org/10.1007/978-3-031-28978-1_17

the Baikal region together with Irkutsk Oblast [1]. During large-scale industrialization in the area under consideration, there was an extensive development of industrial production; there was an influx of population and the growth of cities and towns. The basis of the region's economy – fuel, energy, mining, metallurgical, forestry, pulp and paper industries and mechanical engineering was formed. However, the lack of processing in the basic industries exacerbated the structural imbalances in the national economic complex and contributed to its technological backwardness. All this was most painfully manifested during the transition to the market and continues to hold back the economic revival in some industries until today (engineering and metalworking).

The problems and prospects of the industrial development of Russia, including its eastern regions, are widely studied by domestic scientists [2–7]. Researchers draw attention to the need to accelerate industrial development through the creation of new high-tech industries for the deep processing of raw materials [8–14]. The political and economic events of recent years have become the reason for the reorientation of the foreign economic activity of the Russian Federation to the East. The works on this issue consider the strategic partnership with China [15, 16] and the development of the border areas of the Asian part of Russia [17–19]. All this became the purpose of this study - to determine the geographical differences in the interdependence of the production potential and foreign economic relations of the Eastern Siberian regions within the framework of the eastern vector of development.

2 Materials and Methods

The research is based on the 'Industrial' element of the reindustrialization potential of the subjects of Eastern Siberia, because the foreign economic activity of the regions is mainly determined by industrial production. This element includes a number of indicators. The most important are the shares of the region in the all-Russian value of the volume of shipped industrial products and the value of fixed assets of industry; the share of gross value added (GVA) of industrial production in the total GVA of the region, the volume of shipped products by type of activity 'Manufacturing' in the total volume of shipped industrial products and the value of fixed assets of industry in the total value of fixed assets of the region, as well as indicators of renewal of fixed assets. [20]. Quantitative analysis was carried out using the linear scaling method, due to the large variation in the obtained values, based on the data and index indicators of Rosstat for 2010, 2012, 2014, 2016 and 2018. The production potential index (PPI) itself is formed as the arithmetic mean of the normalized values of all indicators.

To assess foreign trade activity and the level of foreign economic security of the studied subjects and Eastern Siberia as a whole, regional economic indicators were used that reflect their dynamics, the level of development and the balance of foreign trade activity [21, 22]; foreign trade turnover; export quota; import quota; foreign trade quota; import-export coverage ratio; and coefficient of balance of foreign trade. All calculations were made on the basis of Rosstat data for 2010 and 2020 [23]. The paper also uses materials from the Economic Information Agency 'Prime' [24] and the Russian Export Center [25].

3 Results and Discussions

Based on the calculations of the production potential, all regions of Eastern Siberia were divided into 3 groups: I – high potential (above 0.700), II – medium (from 0.300–0.699), III – low (below 0.299) (see Fig. 1).

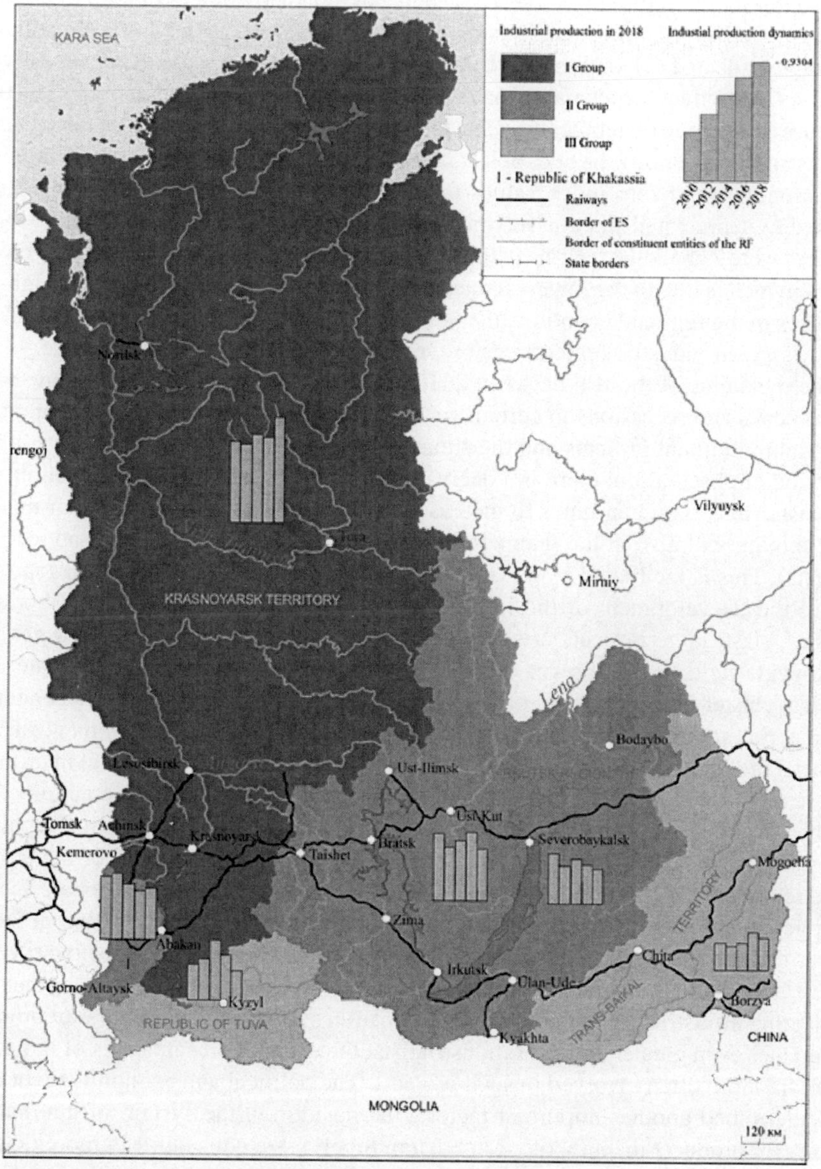

Fig. 1. Production potential of Eastern Siberia. Index: I – high potential (above 0.700), II – medium (from 0.300–0.699), III – low (below 0.299).

Krasnoyarsk Krai has strengthened its leading position in the ranking for the period of observation (from 0.722 in 2010 to 0.930 in 2018). In 2018, the subject had the highest indicators, reflecting the weight of the region's industry in the all-Russian sense, the degree of participation of industry in the structure of the region's economy, and the minimum coefficient for the degree of depreciation of industrial funds (1,000). It should be noted that the region has the most diversified economy. The second group includes Irkutsk Oblast, the Republic of Khakassia and Buryatia with indices of 0.461, 0.458 and 0.304, respectively. During the considered time interval, the production potential for these regions decreased, on average, by 1.3 times. This is largely due to consistently low indicators that reflect the available material resources for the development of industry in the future (the degree of depreciation and the share of completely depreciated fixed assets in industrial production). Irkutsk oblast, despite the rather high indicators of industrial production, has had zero index values for the degree of depreciation and the share of completely depreciated fixed assets in industrial production since 2012. In 2018, the outsiders were Zabaikalskii Krai (0.282) and Tuva (0.274) among the regions of Eastern Siberia, which is due to the lowest indicators in terms of the share of the most high-tech industries in the regional economy, the volume of industrial production and the relative size of assets of industrial enterprises of these subjects in the all-Russian value.

The dynamics of the IPP between 2010 and 2018 has a wave-like structure, where growth peaks and recessions in certain years are associated with the implementation of industrial investment projects and the situation on the world market. In general, during the period of observation, there is a decrease in this potential in the east direction.

Krasnoyarsk Krai continues to increase its production capacity every year not only in the mining, but also in the manufacturing industries (including of high-added value products). This is facilitated by a stable influx of investments in complex megaprojects such as the 'Development of the Lower Angara Region' and the 'Yenisei Siberia'. In Irkutsk Oblast, large projects are also being implemented to supply natural gas from the Kovykta field to the 'Power of Siberia' gas pipeline system, and to create a gas chemical cluster in Ust-Kut and others. The rest of the subjects are most dependent on raw material specialization and lack material resources for the development of other industrial productions. Such large-scale industrial plans with large financial investments in these regions are not yet expected in the near future. In Buryatia, the creation of new and the development of existing industries is complicated by environmental regulation due to the Baikal natural territory (BNT). The BNT occupies over 50% of the territory of the republic, where the main deposits of minerals and forest resources are located. Other geographic factors that affect the development of industrial production in Eastern Siberia include harsh climatic conditions (the northern regions of Krasnoyarsk Krai, Irkutsk Oblast and Zabaikalskii Krai), a rather low level of transport development and engineering infrastructure, which affect the location of production and sales of products. This creates even greater costs for industrialists; therefore, representatives of large businesses are more often involved in such projects. The political and economic pivot to the East determined another important factor - the position in the system of international markets for products manufactured in Eastern Siberia. We consider it in more detail.

Table 1 shows that the obtained values of the foreign trade activity of Eastern Siberia differ significantly from the all-Russian ones. Thus, the indicators of foreign trade, export and import quotas indicate a lower degree of involvement of Eastern Siberia in foreign trade activities in comparison with the Russian Federation. At the same time, the ratio of imports to exports and international competitiveness show that the foreign trade of the considered macroregion is more balanced than that of the Russian Federation as a whole.

Table 1. Indicators for assessing foreign economic interaction of subjects of Eastern Siberia in 2010 and 2020*.

	Foreign trade balance ratio	Export quota	Import quota	Foreign trade quota	Import coverage ratio over exports	Foreign trade turnover (mln USD)
Russian Federation	625980.0 568773.3*	0.27 0.27*	0.16 0.18*	0.21 0.19*	1.73 1.46*	0.27 0.19*
Eastern Siberia	21059.1 20962.9*	0.26 0.22*	0.05 0.07*	0.16 0.14*	4.86 3.29*	0.66 0.53*
Internal Regions						
Krasnoyarsk Krai	11118.5 8905.7*	0.28 0.17*	0.04 0.07*	0.32 0.24*	7.98 2.55*	0.78 0.44*
Irkutsk Oblast	6303.9 7272.2*	0.28 0.28*	0.07 0.07*	0.35 0.37*	4.29 3.70*	0.62 0.57*
Republic of Khakassia	2506.7 1945.0*	0.60 0.44*	0.19 0.08*	0.40 0.26*	3.09 5.30*	0.51 0.68*
Border Regions						
Zabaikalskii Krai	510.0 1519.6*	0.02 0.18*	0.07 0.08*	0.05 0.13*	0.34 2.41*	-0.49 0.41*
Republic of Buryatia	610.8 1232.2*	0.10 0.28*	0.04 0.02*	0.07 0.15*	2.74 16.83*	0.47 0.89*
Republic of Tyva	9.2 88.2*	0.00 0.07*	0.01 0.01*	0.00 0.04*	0.48 8.59*	−0.35 0.79*

Traditionally, border regions, due to their geographical location, actively participate in foreign trade activities with neighboring states and have a more developed economy [18]. The border regions of Eastern Siberia, as well as other regions of the Russian border, have the opposite pattern, which has developed since the days of the USSR. Then the territorial distribution of productive forces in the deep regions of the country was due not only to the presence of a natural resource base, but also to the military-dangerous, barrier nature of the borders [26, p.56]. According to the data obtained, during the period of observation, Zabaikalskii Krai, the Republics of Tyva and Buryatia are less involved in foreign economic relations (Table 1). Thus, the total value of the coefficient of openness of internal regions (Krasnoyarsk Krai, Irkutsk Oblast and the Republic of Khakassia) is 1.2 times higher than the corresponding indicator of border regions.

Interior regions conduct foreign economic activity more efficiently and receive corresponding benefits. They are more export-oriented and therefore export-dependent. However, in recent years, the regions of Transbaikalia show the values of the level close to the internal ones. This is due to the output of mining enterprises of the region (Bystrinskii Mining and Processing Plant, development of Apsatskii and Zashulanskii coal deposits, etc.), the development of the manufacturing industry of the republic and the sale of their products abroad. Based on international standards for the import quota, the economy of all the subjects under consideration is considered sustainable. At the same time, the level of foreign trade self-sufficiency in the border regions has become higher since 2010. In general, the coefficient of international competitiveness for the period under review is balanced for the subjects under study. Zabaikalskii Krai was the exception. It maintained a negative trend in the values of this indicator from 2010 to 2018.

The basis of exports for these two groups of regions is different. Interior regions are characterized by the export of non-energy products of low-added value: aluminum and its alloys, refined copper, ferroalloys, lumber, precious metals, etc. The share of deliveries abroad of primary processing products was 96% in the Republic of Khakassia, 85% in the Krasnoyarsk Territory, Irkutsk region – 62% in 2020. In turn, the border regions are focused on the export of raw materials: coal, ores of ferrous and non-ferrous metals, roundwood (Tuva – more than 99%, Zabaikalskii Krai – 95% and the Republic of Buryatia – 49%). It should be noted that in Buryatia, in the structure of non-commodity non-energy goods, the export of high-tech products accounts for up to 82%. These are mainly aircraft products that are shipped to China, Niger and Mongolia. A significant share of exports of high-added value products in Krasnoyarsk Krai (12% – aerospace and nuclear products) is supplied to the Republic of Korea, Turkey, China and a number of other countries [25]. Since state corporations produce such products, the market will be limited due to the growing sanctions pressure on the Russian economy. The geography of the main trade flows from the subjects under consideration is similar to each other and mainly falls on the countries of the Asia-Pacific Region (China – 60%) and the CIS. Krasnoyarsk Krai is an exception, where exports are mainly oriented to the West [25]. The import structure for internal and border regions, as well as for Russia as a whole, is the same - these are high-tech goods (machinery and equipment) from Europe, the USA, Canada, South Korea and Japan [23].

The position of these two groups of regions in the system of external relations, through the presence of foreign capital, is also different. The domestic ones concentrate more foreign and joint ventures than the border ones. But the dominance of the founders from the PRC is typical for all the subjects under consideration. It should be noted that in Krasnoyarsk Krai and Irkutsk Oblast, the predominance of foreign capital is noted in the timber and woodworking industries (118 and 61 enterprises, respectively). The Zabaikalskii Krai also shows high activity of foreign founders (176 companies). It can be said that China is conducting economic expansion in this territory, where its companies leave 76% of the total number of foreign business enterprises. It is especially abundant in the mining and forestry industries. These are both large holdings and individual entrepreneurs [24]. For this reason, despite the common border, there are no enterprises from Mongolia in the region, as, for example, in the Republics of Buryatia and Tyva. In general, the border regions are characterized by a shortage of investments from European countries, but in the current economic circumstances, this is no longer a problem.

4 Conclusion

Eastern Siberia occupies an important place in the Russian economy due to its natural resource potential and geopolitical position. The territory has sufficient prerequisites for the creation of high-tech industries, but a number of economic and geographical restrictions hinder its development. Renovation of fixed assets and large injections of funds are directed to the extractive industry, while other industries use partially or completely obsolete equipment. According to existing practice, the creation of new industries for the development of natural resources implies a mono-specialization of service settlements with all the ensuing problems. The role of Eastern Siberia as a supplier of raw materials and intermediate products to the world market will only grow stronger. At present, only Krasnoyarsk Krai and Irkutsk Oblast have the resources for reindustrialization. For the sustainable economic development of the regions under study, it is necessary to increase the depth of processing of raw materials in order to receive a significant return in the form of tax revenues to the budget and provide employment for the population, but most often, big business does not see any benefit in this. Unlike the rest of the studied subjects, Krasnoyarsk Krai and Buryatia have a high share of the manufacturing industry in the structure of their economies. If the region is able to develop it both by giant enterprises (the 'Yenisei Siberia' investment project) and by small and medium-sized businesses, then the republic is significantly limited in its capabilities (mono-profile machine building and lack of tools for deploying its new industries).

The border regions of Eastern Siberia use their geo-economic position inefficiently and are inferior to domestic entities in foreign trade activities. The uneven involvement of the SB regions in foreign trade activities is explained by differences in industrial development and the potential of subjects where the border areas have a less complex production structure with a predominance of primary processing of resources. The export of goods and the import of capital in the subjects under study is determined by ties with the PRC. The founders of Chinese business dominate in the macroregion, and the predominance of China is most typical for Transbaikalia, gradually decreasing from east to west. In the current realities, foreign trade relations with the states of Central Asia and the Asia-Pacific region will further strengthen. At the same time, the vulnerability of the economy will only increase due to dependence on the export of raw materials and low-added value products, and imports of high-tech goods. Strengthening interregional ties, as well as with neighboring countries, is possible when creating joint ventures based on industrial parks and clusters. The subjects of Eastern Siberia have enough advantages for successful industrial development, but only with proper use of the existing and activation of hidden potentials.

Acknowledgements. The article was produced based on the materials of the research project of the Irkutsk Scientific Center of the Siberian Branch of the Russian Academy of Sciences "Socio-economic development of a resource region in the context of changing external factors" No. 1210218000157–8 and the RFBR grant "Problem-oriented approach to enhancing interregional interactions" No. 20–010-00990.

References

1. Strategy for the socio-economic development of the Far East and the Baikal region until 2025. https://docs.cntd.ru/document/902195483. Accessed 15 Sep 2022
2. Sysoeva, N.M., Rogovskaya, N.V., Grigorieva, M.A.: A new stage in the industrialization of East Siberia. Geogr. Nat. Resour. **1**(29), 69–73 (2008)
3. Lamin, V.A., Malov, V.Y.: Asian part of Russia: economic development modeling in the context of historical experience. Novosibirsk Publishing House of the SB of the RAS, Novosibirsk (2012)
4. Dets, I.A.: Project approach in russian regional policy through the example of the Baikal region. Regions and Their Socio-Economic Growth. In: Madry, C. (ed.) Bogucky Wydawnictwo Naukowe, pp. 45–54, Poznan (2013)
5. Kuleshov, V.V. (ed.): Present-day Role of Siberian Economy in the Russian Macroeconomic Complex. IEIE SB RAS, Novosibirsk (2014)
6. Ishaev, V., Ivanter, V., Kuvalin, D.: Economy of Far East and Baikal Region: Nation-Wide Line of Approach. MAKS Press, Moscow (2015)
7. Kolomak, E.A. (ed.): Spatial Development of Contemporary Russia: Trends, Factors, Mechanisms. Institutions. IEIE SB RAS, Novosibirsk (2020)
8. Bodrunov, S.D., Grinberg, R.S., Sorokin, D.E.: Reindustrialization of the Russian Economy: imperatives, potential. Risks. Economic Revival of Russia **1**(35), 19–49 (2013)
9. Kuleshov, V.V., Seliverstov, V.E.: Program for reindustrialization of the economy of novosibirsk oblast: ideology and main directions for its implementation. Region: Economics and Sociology **3**(87), 88–122 (2015)
10. Kuleshov, V.V. (ed.): Mineral Resource Sector of Asian Russia: How to Ensure Socio-Economic Returns. IEIE SB RAS, Novosibirsk (2015)
11. Nikolsky, A.F.: "Novyi angarstroi" as a key project for the future industrialization in Russia. Geogr. Nat. Resour. **4**, 143–153 (2017)
12. Tarakanov, M.A.: Problems of manufacturing products for final consumption in the industry of irkutsk oblast. Geogr. Nat. Resour. **39**(2), 168–174 (2018)
13. Bezrukov, L.A.: Territorial structure of the economy and population of the Baikal Region: features. Trends and Prospects. ECO **5**, 8–26 (2019)
14. Rudneva, V.A.: Assessing the reindustrialization potential of regions of eastern Siberia. Geogr. Nat. Resour. **4**, 176–183 (2019)
15. Jhang, C., Yudina, T.N.: Industrial cooperation between Russia and China: opportunities and priorities (within the context of the RF – PRC strategic partnership and XIX CPC congress decisions). Theoretical Economy **2**, 13–21 (2018)
16. Sukhodolov, Y.A.: Modern specifics and prospects of development of foreign trade cooperation between the Irkutsk region and the People's Republic of China. Russian & Chinese Studies **1**(3), 49–57 (2019)
17. Popkova, L.I., Madra, C., Vardomsky, L.B.: Strategy for the Development of Border Territories: Traditions and Innovations. Kursk State University, Kursk (2018)
18. Sysoeva, N., Rudneva, V.: Russian border regions in the economic corridor of China-Mongolia-Russia. Studies of the Industrial Geography Commission of the Polish Geographical Society **35**(2), 53–63 (2021)
19. Bezrukov, L.A., Fartyshev, A.N.: Development of copper industry in Russia and Mongolia: the problem of access to foreign markets. ECO **6**, 115–133 (2022)
20. Novikova, N.V.: Methodology for identifying the potential of new industrial modernization in the space of the macroregion. In: Silin, Y.P., Vlasova, N.Y., Dvoryadkina, E.B. (eds.): Regional economy: challenges, priorities, strategic guidelines, pp. 152–171. Publishing House of the Ural State Economy University, Yekaterinburg (2017)

21. Ultan, S.I.: The analysis of indicators and evaluation methods of foreign trade activities in the region. HERALD OF OMSK UNIVERSITY **3**, 38–49 (2012)
22. Kazitova, E.I.: Indicators of assessment of foreign economic security of region on materials of altay territory. Vestnik Altaiskoy Nauki **2**(24), 166–172 (2015)
23. Regions of Russia. Socio-economic indicators. Federal State Statistics Service (2022). http://www.gks.ru/bgd/regl/b16_14p/Main.htm. Accessed 21 Apr 2022
24. Economic Information Agency "Prime". https://1prime.ru/bir/. Accessed 18 Apr 2022
25. Russian export center. https://www.exportcenter.ru. Accessed 5 May 2022
26. Vardomsky, L.B.: Border belt of Russia: problems and development trends. Russia and the Contemporary World **2**(27), 54–64 (2000)

Check for updates

Formation of Dark-Coniferous Forests Under the Canopy of Light-Coniferous Taiga in Different Conditions of Cisbaikalia

Aleksey I. Shekhovtsov[1(✉)], Aleksander P. Sizykh[2], Aleksander P. Gritsenyuk[2], and Vicktor I. Voronin[2]

[1] V.B. Sochava Institute of Geography, Siberian Branch, Russian Academy of Sciences, Ul. Ulan-Batorskaya 1, 664033 Irkutsk, Russia
ashekhov@irigs.irk.ru

[2] Siberian Institute of Plant Physiology and Biochemistry, Siberian Branch, Russian Academy of Sciences, Ul. Lermontova 132, 664033 Irkutsk, Russia

Abstract. Identification of the structural-dynamic organization of forests of different physical-geographical conditions and characteristics of their formation makes it possible to forecast the possible development trend of forest ecosystems across vast territories. Emphasis is placed on the encroachment of dark-coniferous tree species onto the zone of light-coniferous forests in the form of undergrowth, in the early stages of formation of a forest stand with its encroacvhment on the second growth, which indicates a spatial expansion of the dark-coniferous component of the taiga structure against a background of the climate dynamics over the last several decades in the region. Determination of the trends in forest formation at the regional topological level of organization of the natural environment contributes to determining the genesis of vegetation in general against a background of the modern vector of climate variability. Every so often the undergrowth of trees comprising a dark-coniferous taiga (Siberian stone pine, spruce, and fir) occurs under the canopy of a small-leaved forest stand. Successions of forest-forming species of trees in taiga geosystems from light-coniferou and dark-coniferous components provide express information on the development of forests across time and space. In many respects, such forests can serve as indicators of variability in natural environment at the regional topological level of its organization, which, in turn, is a stabilizing factor for stability of the functioning of forest vegetation in different areas of Cisbaikalia and the Baikal region as a whole.

Keywords: Physical-geographical conditions · Dark-coniferous forests · Light-coniferous taiga · Cisbaikalia

1 Introduction

The physical-geographical conditions of a territory, and the zonal and the altitudinal-belt differentiation of environmental parameters determine the vector of formation of natural systems of any hierarchy. The cyclicity and dynamics of climate that initiate

I. N. Vladimirov et al. (Eds.): RERSDNA 2022, SPEES, pp. 160–166, 2023.
https://doi.org/10.1007/978-3-031-28978-1_18

the possible displacement of natural zones or altitudinal belts as particular environments contribute to the structural differentiation of vegetation witin the boundaries of a geome as a whole. Determination of the consequences of such changes is the prolem of modern biogeography, biogeocenology and ecology as systems of knowledge of the formation and development of vegetation on territories of particular physical-geographical conditions.

The spatial structure of natural systems (geosystems), including forests of any typology, can serve as relatively correct models for purposes of determining the whole of dversity and multidirectionality of changes in the geome against a background of dynamic tendencies of climate. Objects of such studies can be represented by forests of different physical-geographical conditions where the pecsuliarities and specific character of the structural-dynamic organization of forests are clearly pronounced within the boundaries of particuar natural zones or altitudinal belts. Identificaton of the peculiarities of spatio-temporal variability in forest structure is especially relevant in the context of modern trends in climate variability. In this connection, it is necessary to obtain a great body of information (both quantitative and qualitative indicators) on forests for forecasting the possible development trends in forest ecosystems across vast territories.

The goal of this study is to identify the pecsuliarities o the cenotic structure and dynamicss of forest regeeration, and trends in their formation in different areas of Cisbaikalia for different types of their habitats (ecotopes). The key sites of studies were the middle part of the Tunka depression (the left side, foothills, and surroundings of the settlement of Zhemchug); the basin of the Pereemnaya river (its lower reaches), and the basin of the Olkha river (its upper reaches) of different physical-geographical conditions of occurrence of forests in Cisbaikalia.

2 Objects and Methods

The investigations were made by using the method of geobotanical field surveying [1–3] with descriptions of vegetation communities of different years and growing periods. Use was also made of data on the state of forest comunities from previous years (archival material, publications, and maps). Identification of the species composition of vasculr plants in communities used the exiting determinants of Siberia's flora [4–6], while mosses were specified by using L.V. Bardunov's publication [7]. The schematic map shows the study areas – key sites (see Fig. 1).

Data from the publications just cited will serve as baseline information for monitoring the possible changes in the structural-dynamic communities at the contact of environments under climate change. The vegetation communities in the contact zone of the altitudinal belts serve as indicators of the past, ongoingr and possible future changes in the structure of vegetation in this region and can serve as models in forecasting the structural-dynamic organization of vegetation cover in other areas of Cisbaikalia.

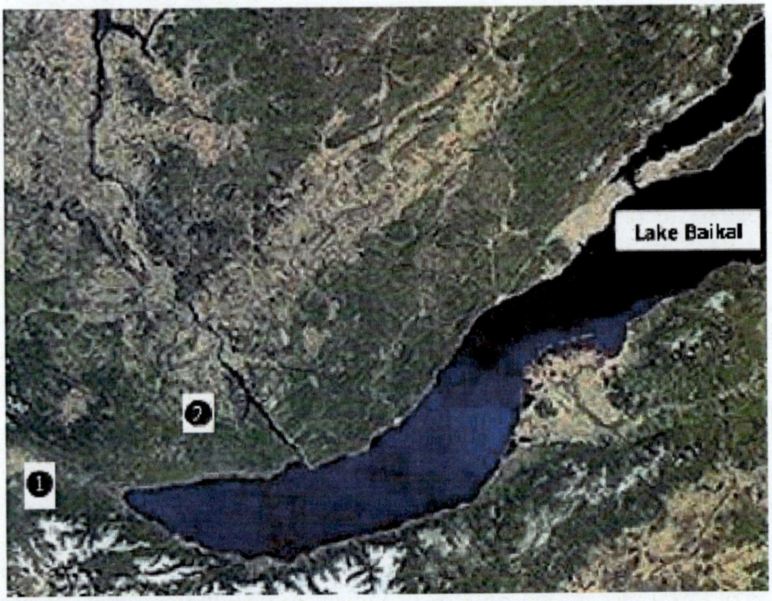

Fig. 1. Schematic map and satellite image (inset), showing the key sites of studies: 1 – Tunka depression, and suburbs of the settlement of Zhemchug; 2 – upper reach of the Olkha river.

3 Results and Discussions

The basis of vegetation in the study area – the bottom of the Tunka depression (its middle part) and its immediate surroundings (see Fig. 1. Point A-1; 51°40′38.34″N, 102°22′33.48″E) consists of larch and pine-larch reedgrass-forbs forests combined with steppizated grass larch stands and areas of (extrazonal) steppes [8]. According to the ecologo-phytocenotic correlation map, vegetation communities are characterized as low-mountain pine, largely subshrup-grass, in places steppizated forestes with fragments of (extrazonal) steppes in combination with green-moss types of forest phytocenoses [9].

Vegetation of the key site is composed of light-coniferous, frequently mono-dominant, pine (*Pinus sylvestris* L.) uneven-aged, with inclusion of *Betula pendula* Roth and *Populus tremula* L., forbs-sedge (*Carex macroura* Meinsh.) forests. The second layer is dominated by pine (*P. sylvestris* L.) with inclusion of small-leaved tree species and Siberian stone pine (*P. pumila* (Pallas) Regel). Larch (*Larix sibirica* Ledeb.) in such forests is represented by small groups or single trees in the composition of communities. The structure of forests in this area of the Tunka depression is characterized by the fact that a considerable role in the composition of phytocenoses is played by repesentatives of the dark-coniferous taiga: Siberian stone pine (*Pinus sibirica* Du Tour) and spruce (*Picea obovata* Ledeb.) in the stages of young growth (tis stage of tree stand) and in the undergrowt of uneven-aged composition: from 2–3-year-old trees to sensually located groups of 10–15-year-old trees. Ground cover of forests is often dominated by sedge (*C. macroura* Miensh.) with a presence of mosses (*Abietinella abietina* (Turn.) Fleisch., *Polytrichum commune* Hedw., *Dicranum polysetum* Sw., and *Rhytidium rugosum* (Hedw.) Kindb.)), characteristic for the popydominant dark-coniferous taiga of zonal type (see Fig. 2).

Fig. 2. Light-coniferous (*Pinus sylvestris* L., *Larix sibirica* Ledeb.) forests of different ages with Asian white birch (*Betula plathyhylla* Sukacz.), undershrubs and green mosses. The undergrowth is dominated by dark-coniferous (*Pinus sibirica* Du Tour, *Picea obovata* Ledeb.) trees species everywere. The undergrowth contains unevenly-aged Siberian stone pine (*Pinus sibirica* Du Tour).

Vegetation of the study area within the Pereemnaya river basin and its immediate surroundings (see Fig. 1. Point A-2; 51°34′42.37″N, 105°18′34.44″E) refers to larch and birch subshrub-green moss recovery sequences of Siberian stone pine, with inclusion of Spruce and larch, ledum-cowberry-green moss, often in cmbination with bergenia Siberian stone pine stands, forests [8]. According to the ecologo-phytocenotic correlation map, the vegetation communities are represented by mid-mountain, largely fir-Siberian stone pine bilberry-small grass-green moss, Siberian stone pine and Siberian stone pine-spruce subshrub-green moss forests and their pine-larch and birch-aspen recovery sequences [9].

Vegetation of the key site is represenetd by mixed light-coniferous, with the upper layer dominated by pine (*P. sylvesteris* L.) and co-dominated by larch (*L. sibirica* Ledeb.), with rare inclusion of spruce (*P. obovata* Ledeb.) and Siberian stone pine (*P. sibirica* Du Tour), green moss (*Polytrichum juniperinum* Hedw., *Dicranum polysetum* Sw., *Aulocomnium palustre* (Hedw.) Schwaegr., *Pleurozium schreberi* (Brid.) Mitt., *Ptilium crista-castrensis* (Hedw.) De Not., *Hylocomium splendens* (Hedw.) B.S.G., and *Climacium dendroides* (Hedw.) Web. et Mohr) forests. Dark-coniferous species of trees are substantially represented in the second layer and in the undergrowth, such as *P. sibirica*

Du Tour, *P. obo*vata Ledeb., and *Abies sibirica* Ledeb. The compositin includes unevenly-aged *B. platyphylla* Sukacz. And *P. tremula* L. The undergrowh of such forests consists of *P. sibirica* Du Tour), *A. sibirica* Ledeb., and *P. obovata* Ledeb. Everywhere. Pine and larch-modern forest-forming species-occur only singly. The layer of subshrubs of these forests is represented by *Vaccinium uliginosum* L., *V. myrtillus* L.), and *V. vitis-idaea* L.), often having a synusial occurrence. Ground cover consists of mosses (*Pleurozium schreberi* (Brid.) Mitt., *Ptilium crista-castrensis* (Hedw.) De Not., *Hylocomium splendens* (Hedw.) B.S.G., and *Climacium dendroides* (Hedw.) Web. et Mohr), and of the edificators o the dark-coniferous zonal taiga of Southern Cisbaikalia. There occurs a synusial presence of *C. macroura* Meinsch.) with taiga forbs: *Pyrola rotundifolia* L., *C. langsdorffii* (Link) Trin., and *Linnaea borealis* L., and lichens of the genus *Cetraria* were observed (see Fig. 3).

Fig. 3. Undergrowth of pine (*Pinus sylvestris* L.) in monodominant pine forests of different age. Soil cover is dominated by sedge (*Carex macroura* Miensh.) / Synusia of mosses (*Abirtinella abie*tina (Turn.) Fleisch., *Polytrichum commune* Hedw., *Dicranum polysetum* Sw., *Rhytidium rugosum* (Hedw.) Kindb.) characteristic for forming of the polydominant dark-coniferous. The undergrowth is dominated by dark-coniferous *Pinus sibirica* Du Tour, and *Picea obovata* Ledeb. Tree species.

Vegetation of the study area in the upper reach of the Olkha river and its immediate surroundings of the key site (see Fig. 1. Point A-3; 51°59′26.99″N, 103°59′19.57″E) refers to pine and larch-pine foxberry-forbs, often with ledum and bog bilberry with the udergrowth of *Duschekia fruticosa* (Rupr.) Pouzar), forests and their aspen-birch recovery sequences [8]. According to the ecologo-phytocenotic correlation map, vegetation communities refer to low- and mid-mountain pine and larch-pine, largely stably-derivative subshrub-grass-green moss forests with fragments of primary dark-coniferous stands [9].

Communities of this key site are represented by *P. sylvestris* L. with *L. sibirica* Ledeb. With inclusion of *P. sibir*ica Du Tour, *P. obovata* Ledeb., *B. plathyhylla* Sukacz. And *P. tremula* L., unevenly-aged *Vaccinium vitis-idaea* L. - *C. macroura* Meinsch. - *Hylocomium splendens* (Hedw.) B.S.G., *Climacium dendroides* (Hedw.) Web. et Mohr, *Pleurozium schreberi* (Brid.) Mitt) forests of the recovery stage of dark-coniferous forests. The second layer includes *P. sylvestris* L. and *L. sibirica* Ledeb., and *P. sibirica* Du Tour. The third layer of the forest stand is dominated by *P. sibirica* Du Tour), *P. obovata* Ledeb. And *A. sibirica* Ledeb. Everywhere, while *P. sylvestris* L. occurs only singly. The undergrowth is dominated by *P. sibirica* Du Tour with inclusion of *A. sibirica* Ledeb. And *P. obovata* Ledeb. Depending on the topographic feature where the quantitative relationships of some or other species vary. Shrubs include *D. fruticosa* (Rupr.) Pouzar, *Ledum palustre* L. and, rarely, *Spiraea media* Fr. Schmidt growing on drier habitats. Subshrubs are represented by *Vaccinium uliginozum* L., *Vaccinium myrtillus* L. and *V. vitis-idaea* L.) of synusial occurrence. Ground cover consists of taiga forbs (Calamagrostis epigeios (L.) Roth, Linnea borealis L., Lycopodium clavatum L., and Lycopodium juniperoideum Sw.) and mosses (*Polytrichum juniperinum* Hedw., *Dicranum polysetum* Sw., *Pleurozium schreberi* (Brid.) Mitt., *H. splendens* (Hedw.) B.S.G., and *Climacium dendroides* (Hedw.) Web. et Mohr), and edificators of the polydominant light-coniferous zonal taiga of Cisbaikalia. Burned-out forests from different years show a recovery (progressive succession), with the main layer dominated by *B. plathyhylla* Sukacz. With inclusion of *P. sylvestris* L. and *L. sibirica* Ledeb.

Characteristic for the structural-dynamic organization of forests growing on all territories considered above are the tendencies toward a succession of the forest-forming tree species as a consequence of the secular dynamics of forests under climate change, mainly a change in humidity, over the course of the last several decades. Since these areas of Cisbaikalia, in accordance with the ecological zoning of the Baikal natural territory [10], form part of the regions of intensive nature management, the identified development vector of forests will depend on the degree of impacts of particular forms of forest management on these territories in general.

4 Conclusions

Characteristic for vegetation of the key site: the vicinities of the settlement of Zhemchug (Tunka depression, Southwestern Cisbaikalia, see Fig. 1, A-1) is the recovery of pine forests on the site of the former arable land and territories which were used previously as pastures for a long time. The presence of representatives of the dark-coniferous taiga in the structure of Siberian stone pine and spruce communities indicates that currently there is taking place a secular dynamics of forests where the forest-forming tree species with the light-coniferous component are substituted by the dark-coniferous component. It should be noted in this connection that such processes were observed for the zonal taiga of the Western-Cisbaikalia type everywhere.

Vegetation of the key site – the upper reaches of the Olkha river (Southwestern Cisbaikalia, see Fig. 1, A-2) is represented by the pine family involving unevenly-aged larch, Siberian stone pine, spruce and fir n layers. The young grouth and undergrowth are dominated by dark-coniferous tree species which appears to be due to the climate dynamics over the last several decades (an increase in humidity) when the edificators of the dark-coniferous taiga – Siberian stone pine, spruce and fir become dominant in light-coniferous forests. These tendencies are also characterized by the dark-coniferous tree species becoming dominant in the undergrowth. Conceivably in this case we observed a succession of forest-forming tree species in forests (the secular dynamics of the zonal taiga) against a background of the climate variability over the last several decades.

The overarching conclusion of this research is the fact that signifgicant processes in the formation of forest vegetation are taking place, which characterize a succession of forest-formin tree species in the areas of Cisbaikalia differing in physical-geographical conditions. It is relevant to poit out here that such a situation with the current trends in forest evolution has been recorded for the Baikal region in general. The modern vector of evelopment of forests in these areas can persist subject to a substantial decrease in the effects of anthropogenic factors: forest cutting of any forms and fires everywhere.

Acknowledgments. The research was done by financial support of the grant no. 075–15-2020–787 of the Ministry of Science and Higher Education of RF for implementation of large scientific projects of priority directions of scientific and technological development (project title "Bases, methods and technologies for digital monitoring and forecasting of the ecological situation on the Baikal natural territory") and funds of a state assignment (theme no. AAAA-A17–117041910172-4).

References

1. Lavrenko, E.M., Korchagin, A.A.: Field Geobotany. Moscow-Leningrad (1964)
2. Bogolubov, A.S.: Methods for Geobotanical Studies. Association "Ecosystem" Publishing House, Moscow (1996)
3. Bogolubov, A.S.: Methodology of Geobotanical Mapping and Profiling. Association "Ecosystem" Publishing House, Moscow (1976)
4. Malyschev, L.I., Peschkova, G.A.: Flora of Central Siberia. Nauka, Novosibirsk, V. 1–2, 1047 p. (1979)
5. Malyschev, L.I., Peschkova, G.A., Polozhij, A.V., Krasnoborov, I.M.: Flora of Siberia. Nauka, Novosibirsk, **1–14** (1987–2003)
6. Baikov, K.S.: Summary of the Flora of Asian Russia. Vascular plants. Nauka, Novosibirsk (2012)
7. Bardunov, L.V.: Key of Leafy Mosses of Central Siberia. Nauka, Novosibirsk (1969)
8. Sochava, V.B.: Vegetation in the south of east Siberia (map, scale 1:1 500 000). State Department of Geodesy and Cartography, Moscow, 4 sheets (1972)
9. Sochava, V.B., Bayborodin, V.N.: Correlational ecological-phytocoenotic map (scale 1:7 500 000). – Irkutsk, 1 sheet (1977)
10. Antipov, A.N.: Ecological zoning of Baikal Natural Territory (map, scale 1:1 000 000). Institute of Geography of RAS SB Publishing House, Irkutsk, 2 sheets (2002)

Interannual Variability of Intrusions and Their Association with Changes in Atmospheric Circulation During 1993–2021

Elena S. Troitskaya(✉) and Mikhail N. Shimaraev

Siberian Branch, Limnological Institute, Russian Academy of Sciences, Ul. Ulan-Batorskaya, 3, 664033 Irkutsk, Russia
elena.troitskaya@lin.irk.ru

Abstract. Water temperature measurements throughout Baikal from the period 1993–2021 were used in investigating interannual variability in characteristics of surface water intrusions and their relationshp with separate mechanisms of atmospheric circulation. The study revealed the spatial nonuniformity of generation of intrusions whose generation frequency decreases from south to north of the lake. The "heat deficit" and the thickness of the renewed layer are comparable in their values in the southern and middle basins and are twice as small in the northern basin of Baikal. Intrusion is most often generated above the western, steeper underwater slope in all basins. Singly occurring intrusions were observed in areas along the eastern shore of South and North Baikal where the gradients of the underwater slope are comparable with the gradients of the western underwater slope. It was found that intrusions in South Baikal in the pre-freezing period are generated with an enhancement of the meridional transfer of air masses and blocking of their zonal transfer. In middle Baikal, owing to its geographical location, May-June intrusions are generated both with an enhancement of the zonal transfer and with an enhancement of the meridional transfer. No significant correlatons with circulation indices were detected for North Baikal.

Keywords: Baikal · Deep convection · Surface water intrusion · Teleconnection indices

1 Introduction

Deep convection arises due to a thermobaric instability of the water column. Usually, this occurs with a change of the water stratification prior to and after freeze-up in deep lakes of temperate latitudes [1]. Deep convection contributes to generation of surface water intrusions into deep or bottom layers of water. As a result, they undergo renewal and saturation with oxygen and biogeni elements, and favorable conditions are maintained for the life of hydrobionts in the deep zone. Hence, deep convection is one of the most important mechanisms in the ecosystem of deep lakes of temperate latitudes.

The phenomenon of deep convection was detected in the Canadian lakes Kamloops (143 m) and Babine (186 m) in the period prior to freeze-up [2], in Lake Crater (USA, 594 m) [3], and in the Japanese lakes Towada (326.8 m) and Shikotsu (363 m) [4].

I. N. Vladimirov et al. (Eds.): RERSDNA 2022, SPEES, pp. 167–174, 2023.
https://doi.org/10.1007/978-3-031-28978-1_19

G.Yu. Vereshchagin [5] was the first to explain the low (3.1–3.4 °C) water temperatures at the bottom of Baikal in terms of the lowering of the cold water from the upper layers due to a decrease in temperature of the maximum density of water with depth (0.21 °C per 100 m). Later [6–10], it was shown that in conditions of a decrease in temperature of maximum density with increasing depth / pressure, in May-June after ice clearance and in November-January before ice setting, in the upper water layer (0–300 m) there can arise a thermobaric instability. It gives rise to deep convection and leads to the lowering of intrusions of cold surface water into the deep and bottom waters. The instability can be initiated by the inhomogeneity of the atmospheric pressure field and by winds in the near-shore zone causing downwellings, and by an increase in water density on the thermal bar front [11].

In Baikal, deep convection reaches a large magnitude, causing intrusions of water with a volume of up to 200 km^3 into the bottom zone of the lake [8] and renewal of the water deeper than 250 m in a year: 10% in South and Middle Baikal, and 15% in North Baikal [12]. Intrusions are responsible for the water aeration in the deep zone [13]. Incoming oxygen support the life of the benthic fauna as well as participating in geochemical processes occurring in the upper layer of bottom sediments [14], and in the decomposition of gas hydrates coming up from the bottom of the lake [15].

The goal of this paper was to study the interannual and spatial variability in characteristics of intrusions caused by deep convection after and prior to ice setting, and their associating with the atmospheric circulation over Baikal.

2 Material and Methods

2.1 Material

This study is based on using data of expedition measurements of water temperature T (±0.002 °C) and oxygen (±0.01 mg/L) with the SBE-25 probe in March – June and, more rarely, in August – September and October during 1993–2021. In 1993, the measurements were made in areas of maximum depths in Baikal. In the other years, the measurements were made at 22 stations of longitudinal section along the thalweg of the lake and 10 transverse sections in South (24–48 stations), 10–11 in Middle (18–36 stations) and North Baikal (20–38 stations, see Fig. 1). In the period of open water in 1993 – July of 2002, and since 2018 till the present, the stations on transverse sections were located 3 and 7 km from the shore and in the middle of the section. In August 2002–2017, additional measurements were made at 1, 5 and 10 km from the shores. In the period of ice setting, in addition to standard stations, measurements were made at 4.5 km from Listvennichnyi Cape and at 3.5 km from Ivanovskii Cape (South Baikal).

Surface water intrusions caused by deep convection are characterized by decreased water temperature and by increased concentration of dissolved oxygen at the bottom (see Fig. 2). Intrusions were signaled by an increase in absolute value of the vertical gradient of T and O_2 by a factor of up to 6–8. In order to estimate the characteristics of the intrusions (heat deficit $\Delta Q_{(-)}$, and height of the layer of renewed bottom water h), the vertical temperature distribution was analyzed (see Fig. 2). Variations in oxygen concentration were not considered in this study.

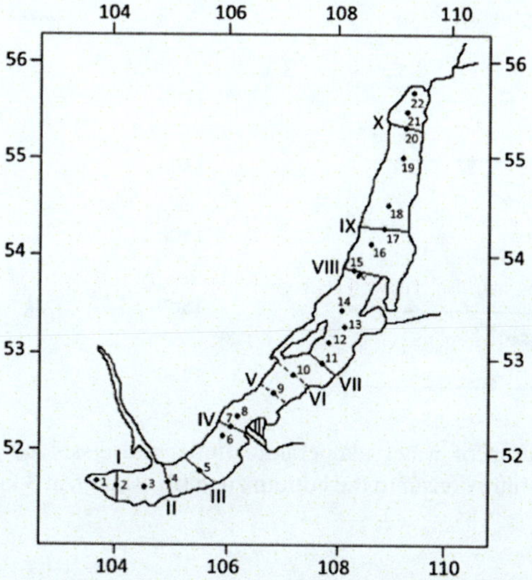

Fig. 1. Schematic map of measurements on the longitudinal (stations 1–22) and transverse sections (I–X). 1 – 15 km from the settlement of Kultuk, central station (c.st.) settlement of Maritui – Solzan river (I), 3 – c.st. Tolstyi Cape – Snezhnaya river, 4 – c.st. Settlement of Listvyanka – settlement of Tankhoi (II), 5 – c.st. Kadil'nyi Cape – Mishikha river (III), 6 – maximum depth opposite Peschanaya Bay, 7 – c.st. Krasnyi Yar Cape – Kharauz by-channel (IV), 8 – minimum depth above the Bugul'deika dike, 9 – c.st. Anga river – Sukhaya river (V), 10 – c.st. Malye Vorota strait – Boldakova river (VI), 11 – c.st. Ukhan Cape – Tonkii Cape (VII), 12 – maximum depth at 7 km SE of Izhimei Cape, 13 – c.st. Khoboi Cape – Krestovskii Cape, 14 – minimum depth above the Academician Ridge, 15 – c.st. Pokoiniki Cape – Orlovskii Cape (VIII), 16 – maximum depth at Zavorotnyi Cape, 17 – c.st. Elokhin Cape – Davshe river (IX), 18 – c.st. Kotel'nikovskii Cape – Amnundakan Cape, 19 – c.st. Muzhinai Cape – Shegnanda Cape, 20 – c.st. Village of Baikal'skoe – Turali Cape (X), 21 – c.st. Tyya – Nemnyanka river, 22 – 7 km from the city of Nizhneangarsk.

2.2 Methods

The thermal effect of cold intrusions or the "heat deficit" $\Delta Q_{(-)}$ (MJ/m^2) in the layer h is calculated as the difference between the actual heat storage of the layer and the hypothetical heat storage in the case of an undisturbed water layer in the bottom region

$$\Delta Q_{(-)} = \rho \cdot C_p \cdot 0.5h \cdot (\Delta T - \Delta T_b). \tag{1}$$

The value of h was determined from the vertical distribution of T in the bottom region (see Fig. 2), ΔT – average measured T in the layer h, ΔT_b – average background T in the layer h for a characteristic distribution of T in the bottom region and in the 200–400-m water column above the renewed water layer. The water density ρ was taken equal to 1000 kg/m^3, and specific heat was taken to be 4.19 kJ/(kg·K).

In order to assess the correlation of the association of intrusions with the atmospheric circulation, we examine the dependencies of $\Delta Q_{(-)}$ with the atmospheric circulation indices *NAO*, *EA*, *EAWR*, *POL*, *SCAND*, *WP* and *PNA* for the period 1993–2021 [16].

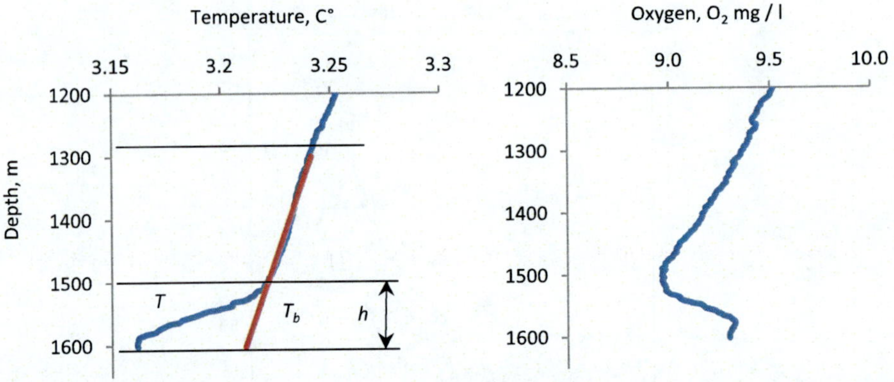

Fig. 2. Vertical distribution of water temperature (left), and dissolved oxygen concentrations (right) for surface water intrusions into the bottom zone (the station at 3 km from Ukhan Cape on 7 June 2021).

3 Results

An analysis of the interannual variability in the characteristics of the intrusions showed that intrusions after the ice clearance are most often generated in South Baikal (66%), in about half of the years under investigation in Middle Baikal (52%), and most rarely in North Baikal (45%). In the pre-freeze-up period and during freeze-up, intrusions in South Baikal were observed only in six out of 29 years (21%). In Middle and North Baikal, no measurements were made in the wintertime, and it is therefore not possible to estimate the frequency of generation of intrusions in these parts of the lake.

The bottom cold layer (*BCL*), formed by intrusions, is most often observed in May – July and decreases before August by a factor of 5, on average. In September, *BCL* almost disappears under the effect of exchange processes. Cold intrusions are of a mass character (*BCL* was recorded at 6–20 stations) as well as of a local character (1–5 stations) in separate years. In 1993, 2002, 2004, 2007, 2009 and 2014, intrusions were recorded in all basins of Baikal, in 1998 and 1999 there were not any intrusions, and intrusions in the other years were produced either in one or two basins of Baikal.

Data from separate stations were used to obtain average (for the basins) monthly values of the characteristics of the intrusions into the bottom zone (see Table 1). They vary over a very wide range (by 2 or 3 orders of magnitude). They are similar for South and Middle Baikal and are by a factor of 1.5–4 smaller for North Baikal.

A cooling of the bottom layer by intrusions in May – July occurred more frequently to reach maximum values of $\Delta Q_{(-)}$ in South and Middle Baikal, whereas the characteristics of *BCL* in North Baikal were decreased (see Fig. 3). We can identify periods when the intesity of intrusions was increased (1993–1999 and 2007–2014) and decreased (2000–2006, and since 2015 till the present). Absolute maxima of $\Delta Q_{(-)}$ correspond to the year 2014 in South Baikal, 2012 in Middle Baikal and 2015 in North Baikal; *h* – to 2000, 2014 and 1997, respectively (see Table 1 and Fig. 3).

For the period under investigation (1993–2021) we identified significant ($p < 0.01$) positive trends in heat deficit (–0.18 (MJ/m^2)/year) and in the thickness of the renewal

Table 1. Mean value and limits (in brackets) of characteristics of cold intrusions, averaged over the basins, as a result of deep convection during 1993–2021.

Basin	Heat deficit, MJ/m^2	Thickness of the renewal layer, m
South Baikal	−6.7 (–0.3 ÷ –26.4)	110 (16 ÷ 188)
Middle Baikal	−7.6 (–0.8 ÷ –15.7)	109 (40 ÷ 186)
North Baikal	−4.9 (–0.2 ÷ –11.4)	82 (26 ÷ 190)

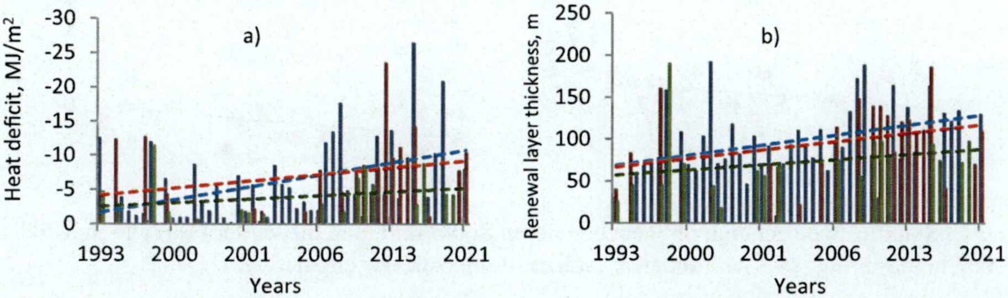

Fig. 3. Average annual values of "heat deficit" (a), and thicknesses of the layer of renewed waters in May – July (b) in South (blue bars), Middle (red bars) and North (green bars) Baikal, and their linear trends (dashed lines of the same color).

layer (1.2 m/year) in South Baikal. For Middle and North Baikal, no significant trends for $\Delta Q_{(-)}$ and h were detected (see Fig. 3).

An analysis of the correlation of the occurrence of intrusions due to deep convection with separate indices of atmospheric circulation showed that the correlation is enhanced with the shift of the averaging of the values of the indices one month earlier where conditions begin to arise for the occurrence of intrusions of surface waters (April–June in the spring, and November–January in the pre-ice-setting period).

We identified significant correlations ($p \leq 0.05$) of the occurrence of intrusions with PNA and EA in the pre-ice-setting period in South Baikal (see Fig. 4a-b) and after ice setting in Middle Baikal – with EA and WP (see Fig. 4c-d). Not any significant correlations were revealed for North Baikal.

4 Discussion and Conclusion

Almost all intrusion events were recorded at stations located in the pelagial and at the western shore of the basins. However, areas near the eastern shore of South Baikal were identified (stations at 7–10 km from the Mishikha river), where intrusions are observed in June–September (8 years out of 29) and in January and March (2 out of 29). One intrusion was recorded at 10 km from the settlement of Murino (South Baikal) and opposite Turali Cape and Davshe river (North Baikal). It seems likely that the occurrence of intrusions along the western shores is due not only to weather conditions but also to the morphometry of the underwater slope. The conditions for the thermobaric instability

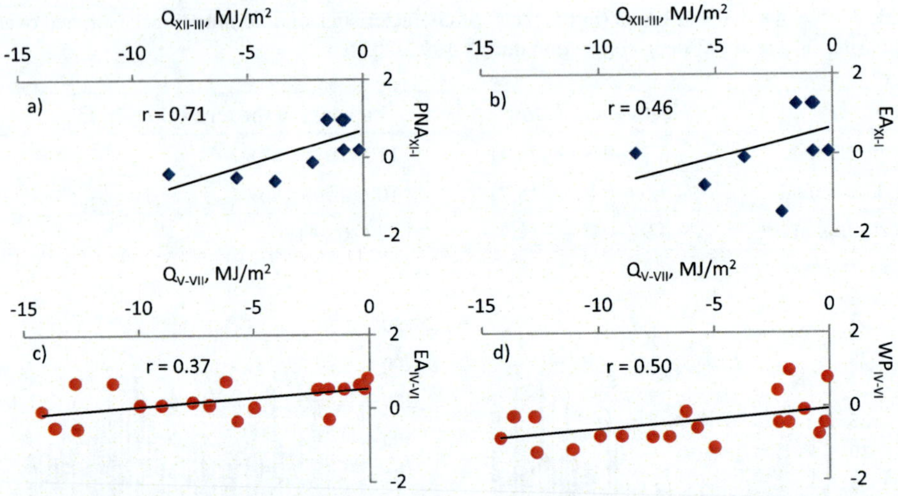

Fig. 4. Significant correlation of "heat deficit" in South Baikal in the winter (a-b) and in Middle Baikal in the spring (c-d) with separate indices of atmospheric circulation.

and, hence, for the intrusion are created above the underwater slopes with large gradients (22–40°) along which there can occur the sinking of the surface waters [9, 10]. In the area opposite the mouth of the Mishikha river and the settlement of Murino at the eastern shore, the gradients are increased (up to 30°) when compared with the prevailing (less than 16°) gradients [17]. Also, in the northern basin, in the areas opposite Turali Cape and Davshe river, the gradient of 24–28° is larger than in the other areas of the eastern shore of North Baikal (up to 18º). In Middle Baikal, intrusions were only observed in the pelagial of the lake with maximum depths, and at the western shore with large gradients of the underwater slopes (28–40º).

Changes in climate as observed since the middle of the 19th century influence the temperature, thermal and ice regimes of the lakes in the northern hemisphere [e.g. 18–20]. For Baikal the influence of climate change has been particularly clearly pronounced since the 1970s. This should also lead to changes in characteristics of the mixing processes which are dependent on water temperature / density. It can be suggested that an increase of the values of "heat deficit" due to surface water intrusions in South Baikal has influenced the temperature regime and heat storage in the botom part of South Baikal. It is shown [21] that the last two decades have exhibited significant trends in a decrease in heat storage in the bottom region in the area of Ivanovskii Cape (South Baikal). However, the absolute value of the trend with respect to the value of heat storage in the bottom zone does not exceed 0.2% and lies in the range of the standard deviation.

The identified correlations of "heat deficit" in South Baikal with the atmospheric circulation indices (see Fig. 4a-b) showed that the intrusions in this part of the lake in the pre-ice-setting period are generated due to the blocking of the zonal transfer of air masses to the eastern part of Eurasia from the north of the Pacific (PNA) and to an enhancement of the meridional transfer of air masses from the eastern Atlantic (EA) over Eurasia.

Intrusions in Middle Baikal after destruction of the ice cover are generated when there occurs either an enhancement of the meridional transfer of air masses from the eastern Atlantic (EA) or with an enhanecement of the zonal transfer from the western part of the Pacific (WP). These mechanisms of the atmospheric circulation in the northern hemisphere are in antiphase but both exert influence on the creation of conditions for occurrence of deep convection in Middle Baikal and for generation of surface water intrusions into the bottom region.

Thus we identified the temporal and spatial inhomogeneity of generation of intrusions. Most often they are observed in South Baikal, and more rarely in Middle and North Baikal. The "heat deficit" and the thickness of the renewed layer are comparable in their values in the southern and middle basins and are twice as small in the northern basin.

It is shown that intrusons, initiated by deep convection, occur not only above the western, steeper underwater slope but also in some areas at the eastern shore of South and North Baikal where the gradients of the underwatr slope are comparable with the gradients at the western shore.

In spite of the identified positive trend in the increase of "heat deficit" in the bottom region, intrusions do not make any substantial contribution to long-term changes in heat storage of the bottom zone.

It is shown that intrusions in South Baikal in the pre-ice-setting period are generated in conditions of the blocking of the zonal transfer and an enhancement in the meridional transfer of air masses. In Middle Baikal, intrusions after the ice clearance can be generated both with an enhancement in the zonal transfer and with an enhancement in the meridional transfer. These effects are associated with the geographical location of the southern and middle basins of the lake.

Multivariate analysis is needed for a detailed study of the correlation of intrusions with the atmospheric circulation. Furthermore, it is necessary to study the contribution of all the currently known processes of deep water renewal: water exchange between basins [12], runoff of tributaries, intrusions due to deep convection [6–10] and thermal bar [11], emergence of gas hydrates of methane [22].

Acknowledgments. We are grateful to the staff members of the Laboratory of Hydrology and Hydrophysics, LIN SB RAS: R.Yu. Gnatovskii and V.V. Blinov for the organization and conduct of expeditionary work, and to N.G. Granin, I.A. Aslamov and V.N. Sinyukovich for helpful remarks and discussion of research results.

This research was performed within the framework of state theme 0279–2021-0004.

References

1. Bouffard, D., Wuest, A.: Convection in lakes. Annu. Rev. Fluid Mech. **51**, 189–215 (2019)
2. Farmer, D.M., Carmack, E.C.: Wind mixing and restratification in a lake near the temperature of maximum density. J. Phys. Oceanogr. **11**(11), 1516–1533 (1981)
3. Crawford, G.B., Collier, R.W.: Long-term observations of deepwater renewal in Crater Lake. Oregon. Hydrobiol. **574**, 47–68 (2007). https://doi.org/10.1007/s10750-006-0345-3

4. Boehrer, B., Fukuyama, R., Chikita, K., et al.: Deepwater stratification in deep caldera lakes Ikeda, Towada, Tazawa, Kuttara, Toya and Shikotsu. Limnology **10**, 17–24 (2009). https://doi.org/10.1007/s10201-008-0257-1
5. Vereshchagin, G.Y.: The main features of the vertical distribution of water mass dynamics in Baikal. In: Academician V.I. Vernadsky to the 50th anniversary of scientific and pedagogical activity, 2, Nauka, Moscow, pp. 1207–1230 (1936)
6. Weiss, R.F., Carmack, E.C., Koropalov, V.M.: Deep-water renewal and biological production in Lake Baikal. Nature **6311**, 665–669 (1991)
7. Shimaraev, M.N., Granin, N.G.: On the question of stratification and the mechanism of convection in Baikal. Dokl. Earth Sci. **321**(2), 381–385 (1991)
8. Wuest, A., Rawens, T.M., Granin, N.G., et al.: Cold intrusion in Lake Baikal: direct observational evidence for deep-water renewal. Limnol. Oceanogr. **50**(1), 184–196 (2005)
9. Schmid, M., Budnev, N.M., Granin, N.G., et al.: Lake Baikal deepwater renewal mystery solved. Geophys Res. Lett. **35**(L09605), 1–5 (2008). https://doi.org/10.1029/2008GL033223
10. Tsimitri, C., Rockel, B., Wüest, A., et al.: Drivers of deep-water renewal events observed over 13 years in the South Basin of Lake Baikal. J. Geophys. Res. Oceans **120**(3), 1508–1526 (2015). https://doi.org/10.1002/2014JC010449
11. Shimaraev, M.N., Granin, N.G., Zhdanov, A.A.: Deep ventilation of Lake Baikal waters due to spring thermal bars. Limnol. Oceanogr. **38**(5), 1068–1072 (1993)
12. Hohmann, R., Kipfer, R., Peeters, F., et al.: Processes of deep-water renewal in Lake Baikal. Limnol. Oceanogr. **42**(5), 841–855 (1997)
13. Shimaraev, M.N., Domysheva, V.M., Gnatovskii, R.Y., et al.: The influence of deep convection on aeration of the bottom zone in Baikal. Geogr Nat Resour **37**, 212–219 (2016). https://doi.org/10.1134/S1875372816030045
14. Pogodaeva, T.V., Lopatina, I.N., Khlystov, O.M.: Background composition of pore waters in Lake Baikal bottom sediments. J. Great Lakes Res **43**, 1030–1043 (2017). https://doi.org/10.1016/j.jglr.2017.09.003
15. Van Rensbergen, P., De Batist, M., Klerkx, J., et al.: Sublacustrine mud volcanoes and methane seeps caused by dissociation of gas hydrates in Lake Baikal. Geol **30**(7), 631–634 (2002). https://doi.org/10.1130/0091-7613(2002)030%3c0631:SMVAMS%3e2.0.CO;2
16. IRI/LDEO Climate Data Library. http://iridl.ldeo.columbia.edu. Accessed 24 Aug 2022
17. LIN SB RAS. http://www.lin.irk.ru/intas/slope.jpg. Accessed 3 Oct 2022
18. O'Reilly, C.M., Sharma, S., Gray, D.K., et al.: Rapid and highly variable warming of lake surface waters around the globe. Geophys Res Lett **42**(24), 10773–10781 (2015). https://doi.org/10.1002/2015GL066235
19. Woolway, R.I., Kraemer, B.M., Lenters, J.D., et al.: Global lake responses to climate change. Nat. Rev. Earth Environ. **1**, 388–403 (2020). https://doi.org/10.1038/s43017-020-0067-5
20. Sharma, S., Richardson, D.C., Woolway, R.I., et al.: Loss of ice cover, shifting phenology, and more extreme events in northern hemisphere lakes. JGR Biogeoscien **126**, e2021JG006348 (2021). https://doi.org/10.1029/2021JG006348
21. Troitskaya, E.S., Budnev, N.M., Shimaraev, M.N.: Changes in the heat content of water column in the slope area of the Southern basin of Lake Baikal in the 21st century. Water **14**(3), 348 (2022). https://doi.org/10.3390/w14030348
22. Granin, N.G., Mizandrontsev, I.B., Kozlov, V.V., et al.: Ring structures on the ice cover of Lake Baikal: analysis of experimental data and mathematical modeling. Geol Geophys **59**(11), 1890–1903 (2018). https://doi.org/10.1016/j.rgg.2018.10.011

Coordinated Degree Assessment on Transport Infrastructure Construction with Regional Economic-Socio-Ecological Development: A Study of the Areas Along the Main Traffic Line in Mongolia

Xu Shuangjie[1,2], Cheng Hao[1,3](✉), Li Fujia[1,3], Zhang Menghan[1,3], Ayana Yangutova[1,3], Tcogto Bazarzhapov[1,3], and GaoYuan[2]

[1] Institute of Geographic Sciences and Natural Resources Research, Chinese Academy of Sciences, 11A, Datun Road, Chaoyang District, Beijing 100101, China
chenghao@igsnrr.ac.cn

[2] College of Geographical Sciences, Liaoning Normal University, DaLian, China

[3] College of Resource and Environment, University of Chinese Academy of Sciences, Beijing, China

Abstract. The China-Mongolia-Russia Economic Corridor is a priority area of the six major economic corridors planned by "the Belt and Road". As a bridge connecting China and Russia, Mongolia's transportation infrastructure construction (TIC) is the key issue to the transportation interconnection among China, Mongolia and Russia. However, the existing research is insufficient to support the actual construction needs. Therefore, this research set the regions along the main north-south traffic line of Mongolia as the study area. We established a regional economic-society-ecology-transportation assessment model (ESET model) to quantitatively assess the construction and development degree of the four subsystems. The results showed that from 2005 to 2015, the development level of the ESET systems gradually increased. The coordination degree between TIC and ESE development has been continuously improved, and most aimags have been optimized as primary coordination. The research results will provide a decision-making basis for the TIC in the China-Mongolia-Russia Economic Corridor, as well as a scientific and technological support for the implementation of the "Belt and Road".

Keywords: Coordinated Degree · Transport Infrastructure Construction · Economic-Socio-Ecological

1 Introduction

The China-Mongolia-Russia Economic Corridor is a priority area of the six major economic corridors of the "Belt and Road". Mongolia as a bridge connecting China and Russia of which TIC is the key issue to China-Mongolia-Russia transportation interconnection and the driving force for regional economic and social development. The lack

I. N. Vladimirov et al. (Eds.): RERSDNA 2022, SPEES, pp. 175–181, 2023.
https://doi.org/10.1007/978-3-031-28978-1_20

of TIC seriously restricts the process of regional development [1, 2]. However, the TIC is closely related to the ESE development of the region [3, 4]. Its construction process is often difficult to coordination the economic, social and ecological (ESE) benefits. The quantitative study of the coordination degree between TIC and regional ESE development is a hot issue of concern to the majority of scholars. In order to solve this problem, some scholars have made useful explorations of the development relationship between ESET.

At present, most of the existing literature is on the development relationship between TIC and economy development [5, 6], urbanization [7, 8], the degree of coordinated development of the ESE system coupling[9, 10]. Chen Guo [11] providing transportation infrastructure investment and economic development is interactive coupling and coordinated interaction. Tie-Ying Liu [12] believes that TIC has a positive effect on urbanization and finds that regionally integrated transport improves the urbanization process. Yu Sun [13] assessed ESE development, and found that economic benefits had the greatest impact on the level of coordinated urban development. Li Qiuyu [14] find the coordination level of tourism-economy-society-ecological environment in Jilin Province was significantly different under the background of global tourism. However, there is still little research on the coordination relationship between the TIC and regional ESE development. In particular, along the main north-south traffic line in Mongolia TIC and regional ESE development is not enough to support the development of its TIC.

In this study, a regional ESET comprehensive system evaluation index system is established, and the entropy value method is used to quantitatively assess the construction and development degree of ESET subsystems. Using the coupling coordination degree model, the degree of coupling coordination between TIC and regional ESE development along the main north-south traffic line of Mongolia is revealed. The research conclusions can promote the sustainable ESE development of Mongolia and neighboring regions, provide scientific support for ensuring the ecological security and sustainable ESE development of the China-Mongolia-Russia Economic Corridor, and have practical application value for promoting the implementation of the "Belt and Road" and the construction of the China-Mongolia-Russia Economic Corridor.

2 Models and Methods

Along the main traffic line of Mongolia the population of accounts for 60.28% of the total population, the GDP accounts for 74.14% of the total GDP. It is an area with a good level of TIC in Mongolia, and an important area along the China-Mongolia-Russia cross-border railway line. The areas are located in the Mongolian plateau in central Asia, which belongs to the arid and semi-arid regions, with scarce water resources and sparse vegetation cover, serious problems of desertification and soil erosion, and fragile ecological and environmental systems. The study area of the TIC and ESE development will threaten its fragile ecological environment. Therefore, this paper takes these six aimags and one city as the object to carry out research on the coordination level of its TIC and regional ESE development (see Fig. 1).

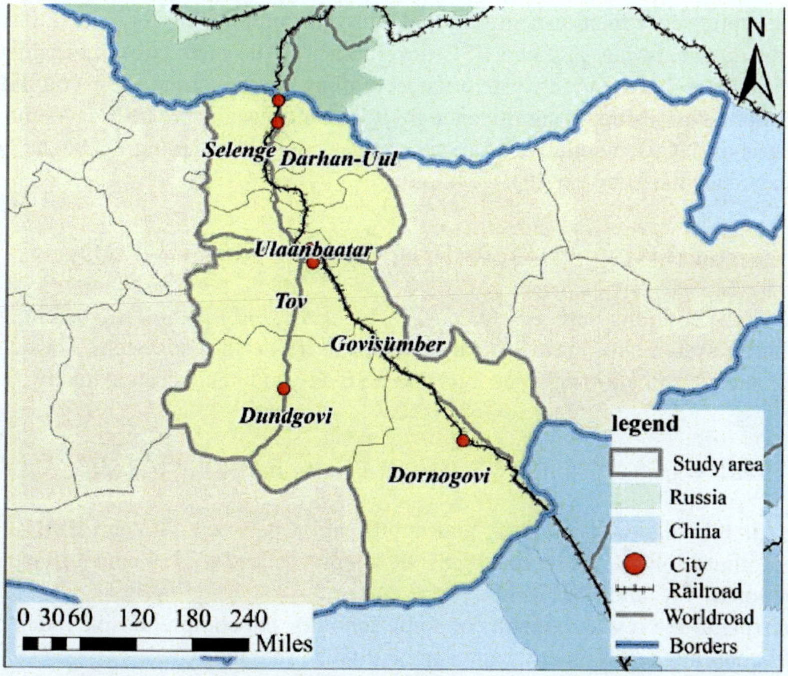

Fig. 1. Map of Mongolia main lines of communication (six aimags and one city).

Based on the quantitative assessment model of regional ESET comprehensive development and the entropy right law, the weights of each system index are determined. The entropy method model is as follows:

$$x'_{\theta ij} = x_{\theta ij}/x_{imax} \quad (1)$$

where $x_{\theta ij}$ is the value of the province i indicator j in θ year, $x'_{\theta ij}$ is the normalized value of the province i indicator j in the year θ, and the x_{imax} is the maximum value of indicator j.

The proportion of the province i indicator j in θ years is calculated:

$$y_{\theta ij} = x'_{\theta ij}/\sum_{\theta}\sum_{i} x'_{\theta ij} \quad (2)$$

The information entropy of the indicator (e_j) is calculated:

$$e_j = -\frac{1}{k}\sum_{\theta}\sum_{i} y_{\theta ij}\ln(y_{\theta ij}),\ k > 0,\ k = \ln(rn) \quad (3)$$

where r is the number of years, n is the target number, and entropy is redundant (g_j), (w_j) and $(H_{\theta j})$ is the indicator weight and score:

$$g_j = 1 - e_j, w_j = \frac{g_j}{\sum_j g_j}, (H_{\theta j}) = \sum_j w_j x'_{\theta ij} \quad (4)$$

The coupling coordination degree model is used to comprehensively assess the degree of coordination between TIC and ESE development. The interaction mechanism and mutual influence between subsystems are coupling systems, and their coupling and coordination relationships is significance for achieving sustainable ESE development in the process of TIC. Quantitative analysis of the degree of coupling between TIC and ESE can be calculated by Eq. (5):

$$C = (f(T) \cdot g(S) \cdot h(E) \cdot i(E))/\left((f(T) \cdot g(S) \cdot h(E) \cdot i(E))/4)^4\right)^{1/4} \quad (5)$$

where C is the coupling degree between TIC and ESE, $f(T)$ is the TIC system, $g(S)$ is the economic system, $h(E)$ is the social system, and $i(E)$ is the ecosystem. The degree of coupling and coordination between TIC and ESE development is calculated by formula (8):

$$T = \alpha \cdot f(T) + \beta \cdot g(S) + \chi \cdot h(E) + \delta \cdot i(E), D = \sqrt{C \times T}, \quad (6)$$

where D is the degree of coupling and coordination between TIC and ESE, T is the degree of impact of TIC on regional ESE development, and α, β, χ and δ illustrate the contribution of TIC and ESE factors to the system.

Referring to the research results of some scholars, the comprehensive index of TIC and ESE coupling development stage can be divided into ten types.

Along the main transport routes in Mongolia with indicators data of TIC and ESE development are from the Statistical Yearbook of Mongolia 2000–2015, the European Space Agency and our group has obtained a large number of first-hand field research data on the ESE after carrying out many joint investigation projects in this area.

3 Results and Discussion

3.1 The Comprehensive Level of TIC Has Been Continuously Improved

From 2005 to 2015, the comprehensive level score of TIC in Mongolia main transportation lines was higher than 0.05, showing an upward trend. The level of TIC in the southeastern Govisumber and Dornogovi has increased significantly. In 2015, the comprehensive score of TIC was higher than that in 2005. It shows that with the proposal of "the Belt and Road" and the economic development of the region, the level of TIC has been continuously improved. In 2015, the highest comprehensive level of TIC was 0.0595 in the Govisumber, and the lowest was 0.0566 in Ulaanbaatar (see Fig. 2).

3.2 The Level of ESE Along the Main Traffic Line in Mongolia Has Improved

From 2005 to 2015, the comprehensive level of ESE systems along Mongolia main traffic line did not change much, and the overall trend showed an upward trend. The ESE integration level of the Govisumber Dornogovi and Tov in the southeast and central region has increased significantly. Among them, in 2005, the comprehensive level scores of the ESE systems of the other five aimags and one city except the Tov were higher than 0.05, and by 2015, the scores were higher than 0.05. It shows that with the level of TIC and the development of industry during the period, the comprehensive development level of the ESE system has become better. In 2015, the highest level of ESE systems was 0.05490 in Selenga, and the lowest was 0.05351 in Ulaanbaatar (see Fig. 3).

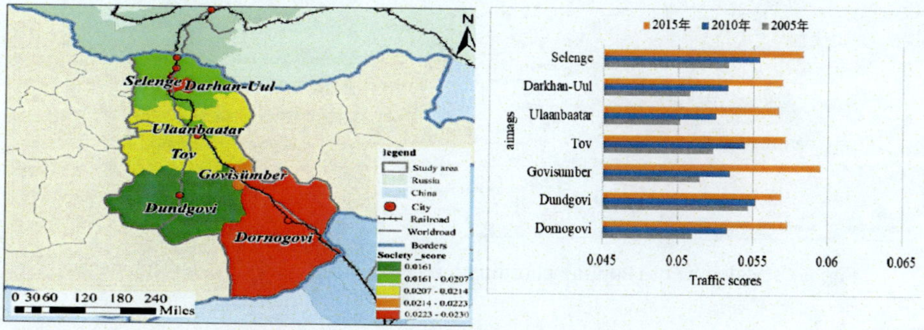

Fig. 2. 2005–2015 Combined scores of TIC.

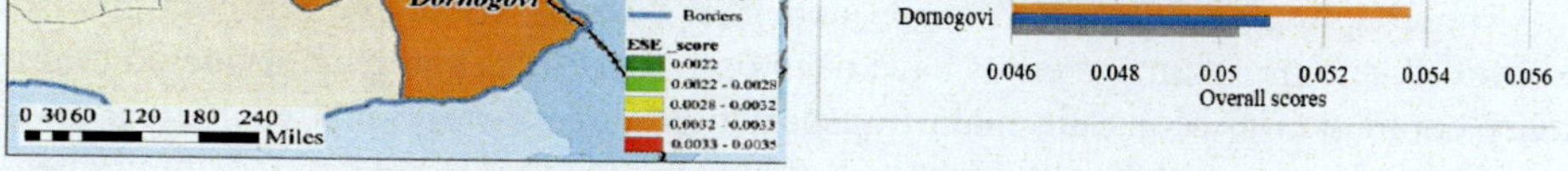

Fig. 3. 2005–2015 Combined scores of ESE systems.

3.3 Coupling Coordination Degree Between TIC and Regional ESE Development

Based on the comprehensive level scores of TIC and regional ESE, the coordination degree of TIC and regional ESE development along the main traffic line in Mongolia from 2005 to 2015 were examined, and the overall trend was on the rise. This shows that the coupling coordination degree of ESE development in the TIC of the study area has been further improved, and the coordination relationship of regional.

In 2015, the degree of coupling and coordination between TIC and ESE development was still barely coordinated, except that the Selenga was still barely coordinated, and the other aimags became primary coordination. The coordination degree of TIC and ESE systems in Mongolia main transportation lines are mostly in the state of primary coordination. The length of the highway and the GDP are the main factors affecting the TIC and the ESE development of the region respectively (see Fig. 4).

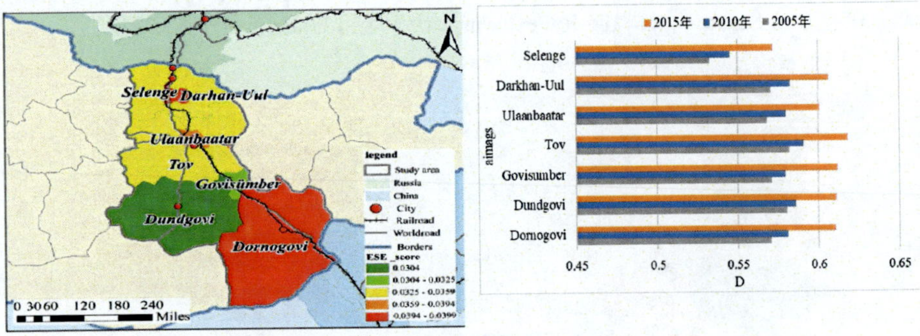

Fig. 4. The degree of coupling coordination between TIC and ESE development.

4 Conclusions

From 2005 to 2015, the coordination degree of TIC and ESE development in the areas along the main transportation lines in Mongolia were quantitatively analyzed. The results show that the degree of comprehensive construction and development of ESET systems in the main transportation areas of Mongolia has gradually increased. The level of ESE integration in the TIC of the Southeast's Govisumber and Dornogovi has been significantly improved.

The coupling coordination of TIC and ESE development has been continuously improved, and most aimags cities have been optimized from barely coordinated to primary coordination. Mongolia main transportation along the regional ecological environment endowment is poor, the pressure of human activities is greater, the comprehensive construction and development of the ESET systems from primary coordination to high-quality coordination still has a long way to go.

In order to achieve an optimal coordination between the TIC and the ESE development, along the main transportation lines of Mongolia should introduce the technology of railway parallel tracks, focus on promoting the construction of railways, rapid highways and high-grade highways, pay special attention to the protection of forests and the conservation of water, and cooperate with China and Russia to build ecological barriers.

Acknowledgements. This research was funded by the National Natural Science Foundation of China (Grant number 32161143029), Key Collaborative Research Program of the Alliance of International Science Organizations (Grant number ANSO-CR-KP-2020–02, ANSO-CR-KP-2020–06), Class A strategic leading science and technology project of Chinese Academy of Sciences (Grant number XDA20030203).

References

1. Dong, S., et al.: Influencing mechanism and policy suggestions of China-Mongolia-Russia high speed railway construction. Acta Geographical Sinica **74**(2), 297–311 (2019)
2. Fattah, M.A., Morshed, S.R., Kafy, A.-A.: Insights into the socio-economic impacts of traffic congestion in the port and industrial areas of Chittagong City, Bangladesh. Transportation Engineering, 100122 (2022)
3. Padma, S., Velmurugan, S., Kalsi, N., Ravinder, K., Erramapalli, M., Kannan, S.: Traffic impact assessment for sustainable development in urban areas. Transportation Research Procedia **48**, 3173–3187 (2020)
4. Qin, C.: Research on transportation infrastructure construction and coordinated socio-economic development. Science and technology economic market, 71–72 (2017)
5. Pradhan, R.P., Bagchi, T.P.: Effect of transportation infrastructure on economic growth in India: the VECM approach. Res. Transp. Econ. **38**, 139–148 (2013)
6. Beyzatlar, M.A., Karacal, M., Yetkiner, H.: Granger-causality between transportation and GDP: a panel data approach. Transportation Research Part A: Policy and Practice **63**, 43–55 (2014)
7. Chen, J., Zhang, W., Song, L., Wang, Y.: The coupling effect between economic development and the urban ecological environment in Shanghai Port. Sci. Total Environ. **841**, 156734 (2022)
8. Pradhan, R.P., Arvin, M.B., Nair, M.: Urbanization, transportation infrastructure, ICT, and economic growth: a temporal causal analysis. Cities **115**, 103213 (2021)
9. Li, L., Fan, Z., Feng, W., Yuxin, C., Keyu, Q.: Coupling coordination degree spatial analysis and driving factor between socio-economic and eco-environment in Northern China. Ecol. Ind. **135**, 108555 (2022)
10. Ding, H., Jiao, X.: Research on coordinated development of economy-ecology-society coupling in "the Blue and Yellow Zone." Journal of China University of Petroleum (Social Sciences Edition) **33**, 24–31 (2017)
11. Chen, G., Chen, Z.: Spatiotemporal coupling coordination between transportation infrastructure investment and economic development in Hubei Province. Railway Transport and Economy **43**, 51–57 (2021)
12. Liu, T.-Y., Su, C.-W.: Is transportation improving urbanization in China? Socioecon. Plann. Sci. **77**, 101034 (2021)
13. Sun, Y., Cui, Y.: Evaluating the coordinated development of economic, social and environmental benefits of urban public transportation infrastructure: case study of four Chinese autonomous municipalities. Transp. Policy **66**, 116–126 (2018)
14. Li, Q., Zhu, L., Wang, J.: Research on the coordination of tourism-economy-society-ecological environment in Jilin province under the background of global tourism. Geographical Sciences **40**, 948–955 (2020)

Check for updates

Spatial-Temporal Evolution of Land Surface Thermal Environment and Its Driving Factors in Baikal Rim Region

Zhang Menghan[1,2], Dong Suocheng[1,2], Cheng Hao[1(✉)], Li Fujia[1,2], Ayana Yangutova[1,2], Tcogto Bazarzhapov[1,2], and Xu Shuangjie[1,3]

[1] Institute of Geographic Sciences and Natural Resources Research, CAS, Beijing, China
chenghao@igsnrr.ac.cn
[2] College of Resource and Environment, University of Chinese Academy of Sciences, Beijing, China
[3] College of Geographical Sciences, Liaoning Normal University, Dalian, China

Abstract. The Baikal region and its surroundings are one of the most unique regions with high ecological conservation value. In recent years, the rapid development of industry and ecotourism, increasing transportation infrastructure, and urban expansion in this region are affecting the natural ecological conditions, including the land surface thermal environment (LSTE). The coordinated economic-social-ecological development in this region has become a hot scientific proposition of great interest to the international academia. In this study, Irkutsk Oblast and the Republic of Buryatia in the Baikal Rim region were selected as the study area, and based on MODIS data in 2011 and 2021, the Google Earth Engine platform was used to invert the land surface temperature (LST), analyses the and characteristics of the LSTE spatial-temporal evolution pattern, and reveal the main driving factors. The results show that the LST in the study area increased from 9.0 °C to 12.2 °C over the last 10 years, with the LST in Irkutsk Oblast increasing more significantly with an average annual growth rate of 2.45%. The spatial pattern of the LSTE shows a spreading pattern from the central city to the surrounding area and along the railway line to both sides. Urbanization development, economic improvement, and transportation infrastructure construction are important driving factors. The research results will provide scientific and technological support for the coordinated development of the ecological environment and economic society in the Baikal Rim region, and also provide a reference for the regional ecological environmental protection and green development in China-Mongolia-Russia Economic Corridor.

Keywords: Land Surface Thermal Environment · Surface Urban Heat Island Intensity · Spatial-Temporal Evolution · Driving Factors · Baikal Rim Region

1 Introduction

In recent years, increasingly intense climate change is affecting the long-term global warming trend more profoundly. Compared to the rest of the world, the Baikal Rim

I. N. Vladimirov et al. (Eds.): RERSDNA 2022, SPEES, pp. 182–188, 2023.
https://doi.org/10.1007/978-3-031-28978-1_21

region in Siberia is the most prominent, and it is one of the three regions with the fastest climate change in the world [1, 2]. Lake Baikal and its surrounding areas not only have unique and complex natural ecosystems but also have a series of economic, social, ecological, and cultural effects [3–5]. In the context of the in-depth development of the China-Mongolia-Russia Economic Corridor, frequent and intense economic and social activities are and will continue to occur in the area around Lake Baikal [6]. These factors are profoundly affecting the natural ecological environment of the Baikal Rim region, including the land surface thermal environment (LSTE).

As a special cross-border area, the Baikal Rim region is not only the frontier area of the China-Mongolia-Russia Economic Corridor and a key area for cooperation between neighboring countries but also has a high value of ecological protection. Therefore, it is very important to monitor and analyze the natural ecological environment in this region. Whether the LSTE is good or not is one of the important indexes to measure the ecological environment [7]. In recent years, scholars from different countries have conducted numerous studies about the measurement by different methods [8, 9], the driving factors [10–12], and the mitigation measures [13].

In the context of the deepening of the China-Mongolia-Russia Economic Corridor, along with the accelerating urbanization and increasing economic development in Russia and Mongolia in recent years, the Baikal Rim region has witnessed a rapid development of industry and ecotourism, gradual expansion of the central cities, and improving transportation infrastructure. These intense anthropogenic activities will become more frequent and intense, which will have a profound impact on the LSTE and heat island effect in the region. As the most sensitive climate regulator in the Northern Hemisphere, changes in the LSTE will greatly influence the feedback mechanisms of regional climate regulation. Therefore, the study of the LSTE in the Baikal Rim region and its driving factors has become a particularly urgent issue.

In this study, Irkutsk Oblast and the Republic of Buryatia, were selected as the study area, and based on MODIS data in 2011 and 2021, the Google Earth Engine platform was used to invert the land surface temperature (LST), analyses the and characteristics of the LSTE spatial-temporal evolution pattern, and reveal the main driving factors. The results will provide scientific and technological support for the coordinated development of the ecology and economy in the Baikal Rim region, and also provide a reference for the regional ecological environmental protection and green development in China-Mongolia-Russia Economic Corridor.

2 Objects and Methods

Lake Baikal is located in the southern part of Eastern Siberia and is situated in the Republic of Buryatia and Irkutsk Oblast. As the closest region to Lake Baikal, Irkutsk Oblast and the Republic of Buryatia (see Fig. 1) in the Baikal Rim region, were selected as the study area.

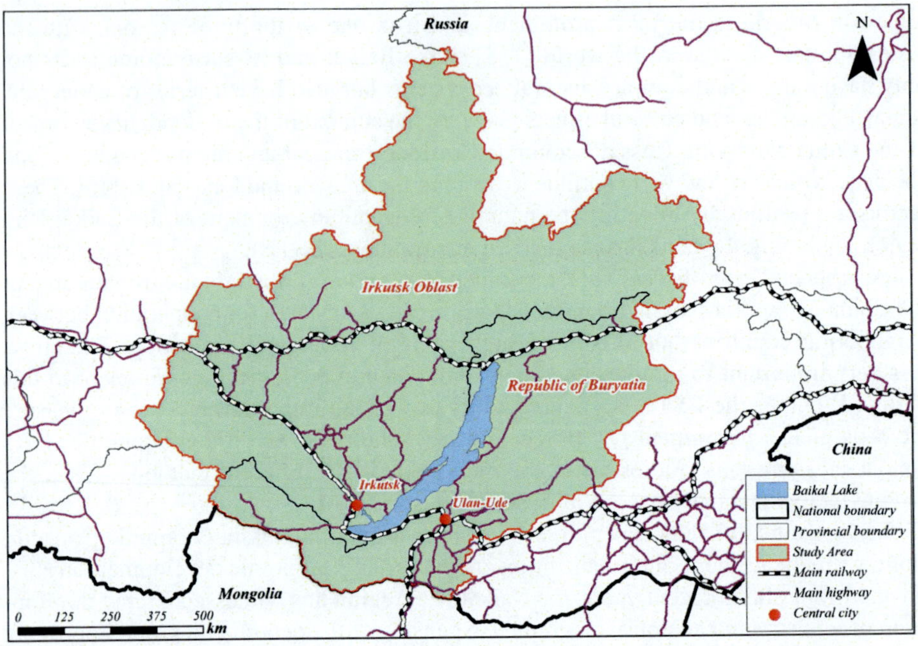

Fig. 1. Location of Baikal Rim region (Irkutsk Oblast and the Republic of Buryatia).

LST data, land cover data, normalized difference vegetation index (NDVI), and nighttime light index data were used in the process of LSTE analysis. The LST data are from MODIS 8-day synthetic average LST products (MOD11A2) for the study area in summer (July-September). Land cover data were obtained from MCD12Q1 products provided by the USGS. The NDVI data are from MODIS monthly NDVI dataset 16-day synthetic product (MOD13A3). The DMPS/OLS nighttime light data is the source of nighttime light remote sensing data for 2011, and the NPP-VIIRS data was selected as the source of nighttime light remote sensing data for 2021.

The research team visited Russia and Mongolia several times and obtained plenty of first-hand research information on the economic, social, and ecological aspects of the study area, which was used as the basic socio-economic data for this study. In addition, some of the data were obtained from the Russian Statistical Yearbook provided by the official website of the State Statistics Service of the Russian Federation.

In this study, we selected MODIS/Terra and MODIS/Aqua daytime and nighttime 8-day synthetic average LST products (MOD11A2) of the study area for the whole years 2011 and 2021, and the image element DN values were converted to LST(°C) by Eq. (1) on Google Earth Engine cloud platform. And then, the average value was calculated to obtain the annual-scale LST with a spatial resolution of 1 km.

$$T_S = DN \times 0,02 - 273,15, \tag{1}$$

where T_S is the LST (°C), and DN is the digital number of each pixel.

Since heat island is a relative temperature distance level of urban over rural, this distance level has no unified threshold standard for either surface urban heat island intensity (SUHI) or meteorological heat island so far. Here, the SUHI and class classification method proposed by Ye et al. (2011) [14] is used to estimate:

$$SUHI_i = T_i - \frac{1}{n}\sum\nolimits_1^n T_{crop}, \tag{2}$$

where $SUHI_i$ is the SUHI (°C) corresponding to the *i*-th pixel on the image, T_i is the LST (°C), n is the number of effective pixels in the rural background, and T_{crop} is the LST (°C) in the rural background.

3 Results and Discussion

3.1 Spatial-Temporal Evolution of LST in Baikal Rim Region

In general, the LST in the Baikal Rim region has increased significantly in the past 10 years. The LST has increased by 3.2°C, with an average annual growth rate of 2.14%. In terms of administrative regions, in 2011, the average LST in the Republic of Buryatia was 9.4 °C. By 2021, the average LST has increased to 10.7 °C, with an average annual growth rate of 1.31%. While the average LST in Irkutsk Oblast has reached 10.1 °C. By 2021, the average LST has increased to 12.9 °C, with an average annual growth rate of 2.45%, which is higher than the average level of the study area.

The spatial-temporal evolution pattern of the LSTE in this region presents three high-value areas in the central-western regions, the north, and the south (see Fig. 2). The high-value areas in the central and western regions are mainly distributed along the main railway lines, while the high-value areas in the north are concentrated in the three border areas – Kirenskiy, Mamsko-Chuyskiy and Kazachinsko-Lenskiy, and the LST has shown a trend of spreading around these three border areas. The high-value areas of the LST in the Republic of Buryatia are concentrated near Ulan-Ude.

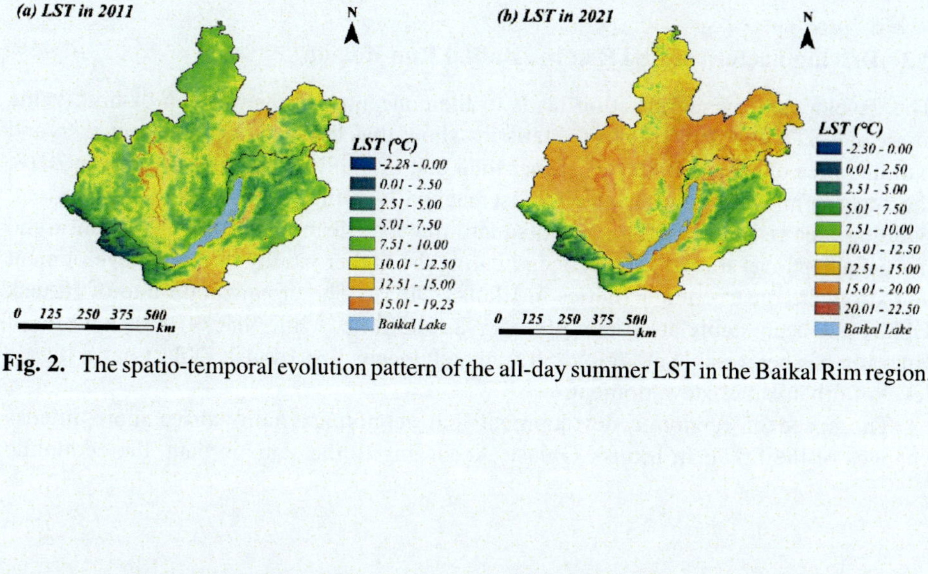

Fig. 2. The spatio-temporal evolution pattern of the all-day summer LST in the Baikal Rim region.

3.2 Spatial-Temporal Evolution of SUHI in the Baikal Rim Region

In general, the SUHI in the Baikal Rim region has remained stable in the past 10 years. In 2011, the SUHI in the Baikal Rim region was −0.5 °C, and by 2021, the SUHI has decreased to −0.6 °C. In terms of administrative regions, in 2011, the SUHI in the Republic of Buryatia was −1.01 °C, and by 2021, the SUHI was reduced to -2.16 °C. While the SUHI of Irkutsk Oblast in 2011 was −0.28 °C, and in 2021, the SUHI has increased to 0.03 °C, a higher increase than the average level of the study area.

The spatial-temporal evolution pattern of the SUHI in this region presented three high-value areas in the western regions, the north, and the south (see Fig. 3). In the past 10 years, the high-value areas in the north were concentrated in the three border areas – Kirenskiy, Mamsko-Chuyskiy, and Kazachinsko-Lenskiy. And the SUHI has shown a trend of extending along the borderline to the northeast and along the main railroad line to the south. While the high-value areas in the western regions were mainly distributed in another three border areas – Chunskiy, Tayshetskiy, and Bratskiy. The high-value areas of the SUHI in the south were concentrated near Ulan-Ude.

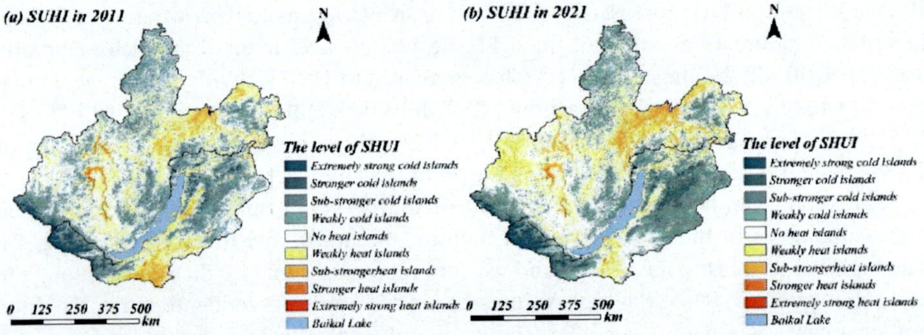

Fig. 3. The spatial-temporal evolution pattern of the all-day summer SUHI in the Baikal Rim region.

3.3 Driving Factors of LSTE in the Baikal Rim Region

The development of urbanization leads to the concentrated distribution of high-value areas of LSTE in Irkutsk Oblast. Statistics show that the total city housing area and building area of Irkutsk Oblast is higher than that of the Republic of Buryatia. In 2018, the total city housing area in Irkutsk Oblast reached 49 million square meters, and the total building area reached 1597 thousand square meters. Meanwhile, this increasing trend has been kept over the past 10 years, indicating the higher vitality of urban development and stronger construction activities in Irkutsk Oblast. The urbanization rate of Irkutsk Oblast has been stable at around 80% for a long time, while that of the Republic of Buryatia has been stable at around 60%, which means that Irkutsk Oblast has a higher level of urbanization development.

The speed of economic development and economic vitality drive more intense changes of the LSTE in Irkutsk Oblast. According to the statistic data, the economic development level of Irkutsk Oblast was higher than the Republic of Buryatia, with an average annual growth rate of 9.43%. The economic vitality of Irkutsk Oblast was higher than that of the Republic of Buryatia. The average annual growth rate of the investment in fixed capital was 11.73%, which was significantly higher than that in the Republic of Buryatia. While the economic vitality of the Republic of Buryatia is relatively low, the investment in fixed capital was stable at around 50 000 million rubles, with an average annual growth rate of only 2.60%.

The construction of transportation infrastructure promotes the linear agglomeration of the LSTE. In the western Irkutsk Oblast in the Baikal Rim region, there was an area with a significantly increased SUHI. In this area and its surroundings, the SUHI has generally shown a trend of spreading from the railway lines to both sides and surrounding vast rural areas and also has a tendency to extend along the main railway lines. These tendencies were mainly due to the supporting projects and electrification construction of the two major railway lines across the Baikal Rim region, the Baikal-Amur Railway and the Siberian Railway. The construction of the two major railway lines has accelerated the social and economic development of the areas along the lines, and further promoted the linear agglomeration of the LSTE. From 2011–2018, the average annual freight shipment by public railway transport in Irkutsk Oblast was 5.3 times that of the Republic of Buryatia, and the passenger departures by public railway transport in Irkutsk Oblast was 6.5 times that in the Republic of Buryatia.

4 Conclusions

The average LST in the Baikal Rim region has increased from 9.0 °C in 2011 to 12.2 °C in 2021. In the past 10 years, the spatial pattern of the LSTE in this region has shown a trend of spreading from the central city to the surrounding area and both sides along the railway line.

The SUHI in the Baikal Rim region has remained stable, decreasing from −0.5 °C in 2011 to −0.6 °C in 2021. The SUHI in this region has three high-value areas in the western border areas, the northern border areas and the south, and it shows a trend of spreading to both sides and surrounding areas along the main railway line.

The urban construction activities accompanied by urbanization have led to the concentrated distribution of high-value areas of the LSTE in Irkutsk Oblast; the speed of economic development and economic vitality have driven more intense changes of the LSTE in Irkutsk Oblast; the supporting projects and electrification construction of the Siberian Railway and the Baikal-Amur Railway have enabled the rapid development, which in turn promotes the linear accumulation of the LSTE.

Acknowledgements. This research was funded by Key Collaborative Research Program of the Alliance of International Science Organizations(Grant number ANSO-CR-KP-2020–02, ANSO-CR-KP-2020–06), Class A strategic leading science and technology project of Chinese Academy of Sciences (Grant number XDA20030203), and National Natural Science Foundation of China (Grant number 32161143029, 42071282).

References

1. Shimaraev, M.N., Kuimova, L.N., Sinyukovich, V.N., Tsekhanovskii, V.V.: Manifestation of global climatic changes in Lake Baikal during the 20th century. Dokl. Earth Sci. **383**(3), 288–291 (2002)
2. Clarke, A., et al.: Climate change and the marine ecosystem of the western Antarctic Peninsula. Philosophical Transactions of the Royal Society. Biological Sciences **362**(1477), 149–166 (2007)
3. Khodzher, T.V., Domysheva, V.M., Sorokovikova, L.M., Tomberg, I.V., Sakirko, M.V.: Hydrochemical studies in Lake Baikal: history and nowadays. Limnology and Freshwater Biology **1**, 2–9 (2018)
4. Panchenko, M.V., Domysheva, V.M., Pestunov, D.A., Sakirko, M.V., Shamrin, A.M., Shmargunov, V.P.: Carbon dioxide in the atmosphere-water system and biogenic elements in the littoral zone of Lake Baikal during period 2004–2018. J. Great Lakes Res. **46**(1), 85–94 (2020)
5. Bilgaev, A., Dong, S., Li, F., Cheng, H., Sadykova, E., Mikheeva, A.: Assessment of the current eco-socio-economic situation of the Baikal region (Russia) from the perspective of the green economy development. Sustainability **12**(9), 3767 (2020)
6. Safronov, A.N.: Effects of climatic warming and wildfires on recent vegetation changes in the Lake Baikal basin. Climate **8**(4), 57 (2020)
7. Yao, Y., Chen, X., Qian, J.: Research progress on the thermal environment of the urban surfaces. Acta Ecol. Sin **38**, 1134–1147 (2018)
8. Chen, F., Yang, S., Yin, K., Chan, P.: Challenges to quantitative applications of Landsat observations for the urban thermal environment. J. Environ. Sci. **59**, 80–88 (2017)
9. Takahashi, K., Yoshida, H., Tanaka, Y., Aotake, N., Wang, F.: Measurement of thermal environment in Kyoto city and its prediction by CFD simulation. Energy and Buildings **36**(8), 771–779 (2004)
10. Yang, J., Yang, Y., Sun, D., Jin, C., Xiao, X.: Influence of urban morphological characteristics on thermal environment. Sustain. Cities Soc. **72**, 103045 (2021)
11. Zhang, M., Dong, S., Cheng, H., Li, F.: Spatio-temporal evolution of urban thermal environment and its driving factors: case study of Nanjing. China. PloS One **16**(5), e0246011 (2021)
12. Yu, Z., Yao, Y., Yang, G., Wang, X., Vejre, H.: Strong contribution of rapid urbanization and urban agglomeration development to regional thermal environment dynamics and evolution. For. Ecol. Manage. **446**, 214–225 (2019)
13. Chen, H., Ooka, R., Huang, H., Tsuchiya, T.: Study on mitigation measures for outdoor thermal environment on present urban blocks in Tokyo using coupled simulation. Build. Environ. **44**(11), 2290–2299 (2009)
14. Ye, C., Liu, Y., Liu, W., Liu, C., Quan, W.: Research on urban surface heat environment monitoring indexes and its application. Meteorological Science Technol. **39**(1), 95–101 (2011)

Study on the Spatial and Temporal of Scientific and Technological Resource Input and Output in Countries Along the Belt and Road

Yan Zhuang[1,2], Fujia Li[1,2], Suocheng Dong[1,2](✉), Qian Liu[1,2], Shuangjie Xu[1,3], Menghan Zhang[1,2], and Kexin Guo[1,2]

[1] Institute of Geographic Sciences and Natural Resources Research, Chinese Academy of Sciences, Beijing 100101, China
dongsc@igsnrr.ac.cn
[2] College of Resource and Environment, University of Chinese Academy of Sciences, Beijing 100049, China
[3] College of Geographical Sciences, Liaoning Normal University, Beijing, China

Abstract. Building a scientific and technological community is an important part of "the Belt and Road" construction. Therefore, a reasonable and objective evaluation of the scientific and technological resources in countries along "the Belt and Road" is of great significance for promoting the interconnection of countries. The time series and Cold-hot spot distribution are used to measure and compare the level of scientific and technological resources in the countries along the "Belt and Road". On this basis, the main input and output elements of the scientific and technological resources of the countries along the "Belt and Road" are analyzed and researched. The results showed that: (1) Different types of science and technology input elements presented different distribution patterns, resulting in spatial differences in the output of science and technology resources. (2) The construction of joint laboratories, cooperation in science and technology parks, and technology transfer would effectively narrow the gap between countries in the output of scientific and technological resources. This study will provide scientific support for promoting the establishment of "Science and Technology Innovation Alliances" in countries along the "Belt and Road".

Keywords: Scientific and technological resources · Input and Output · The Belt and Road · Spatial differences · Scientific support

1 Introduction

All countries in the world will promote the comprehensive development of science and technology as a national strategy. The quantity of scientific and technological resources and the degree of optimization of their allocation and utilization directly affect the realization of the development strategy of science and technology, and will have a profound impact on national scientific and technological innovation activities and economic development [1, 2]. How to optimize the allocation of scientific and technological

I. N. Vladimirov et al. (Eds.): RERSDNA 2022, SPEES, pp. 189–193, 2023.
https://doi.org/10.1007/978-3-031-28978-1_22

resources and continuously improve the allocation efficiency of scientific and technological resources is an important issue facing all countries [3]. The "Belt and Road" requires scientific and technological innovation, and scientific and technological resources are the precursor and support for the innovation and construction of the "Belt and Road". Along with top-level design and practical actions, scholars' research on the scientific and technological resources of the "Belt and Road" is also deepening [4–6]. At present, increasing the investment in scientific and technological resources is not only the focus of academic research, but also a common practice to improve the innovation capacity of the Belt and Road, that is, to promote the improvement of regional innovation capacity by changing the scale and structure of scientific and technological resource investment [7–10]. Investment in scientific and technological resources is a necessary condition for innovation, and the key to whether we can enhance the ability of scientific and technological innovation lies in whether scientific and technological resources can be optimally allocated and fully utilized. The output of scientific and technological resources is the basis of scientific and technological resource cooperation, under the current system, the output of scientific and technological resources directly determines whether countries along the "Belt and Road" can fully carry out scientific and technological resources cooperation [11–15].

Therefore, countries need to make full use of scientific and technological resources to cooperate to promote the construction of the "Belt and Road", use science and technology to help build five "Silk Roads", jointly carry out scientific research, strengthen the exchange of scientific and technological personnel, cultivate scientific and technological talents, and jointly build the "Belt and Road" [16]. Based on this urgent need, this study comprehensively analyzes the overall evolution trend and spatial dynamic distribution pattern of the input and output elements of scientific and technological resources in the major countries along the "Belt and Road" through the use of hot and cold spot analysis methods, so as to conduct in-depth research on the spatio-temporal changes and regional differences of scientific and technological resources of countries along the "Belt and Road", with a view to more rationally allocating effective scientific and technological resources for countries along the "Belt and Road", and truly building the "Belt and Road" provides reference.

2 Models and Methods

The indicators for evaluating the input of scientific and technological resources in this study include the input of human resources, capital resources, and information resources. The indicators for the output of scientific and technological resources include the output of scientific and technological achievements and the output of scientific and technological transformation. The data source is the World Bank.

The Getis-Ord Gi* index was proposed by Getis and Ord, which can find subregions in the study area whose attribute values are significantly different from other places. Based on the Gi* index, it is possible to display the spatially clustered locations of spatially high-value (hot) or low-value (cold) features. The hotspots and coldspots of the input and output of various scientific and technological resources in countries along the Belt and Road can represent their high-level and low-level agglomeration areas, respectively [17–20].

3 Results and Discussion

Based on data acquisition and completeness, this paper selected 2018 to analyze the spatial pattern of scientific and technological resource output of countries along the "Belt and Road".

Total number of scientific and technological papers. The total number of scientific and technological papers published in countries along the "Belt and Road" has increased steadily from 2005 to 2018. After the Chinese government proposed the "Belt and Road" for the first time in 2013, more and more scholars began to pay attention to the scientific problems in the development process of countries along the "Belt and Road" and carried out joint scientific research in cross-border areas. The increasingly strong academic atmosphere contributed to the significant increase in the total number of scientific and technological papers published after 2013. In 2018, the total number of scientific and technological papers increased to about 1.20 million, reaching the highest level (Fig. 1).

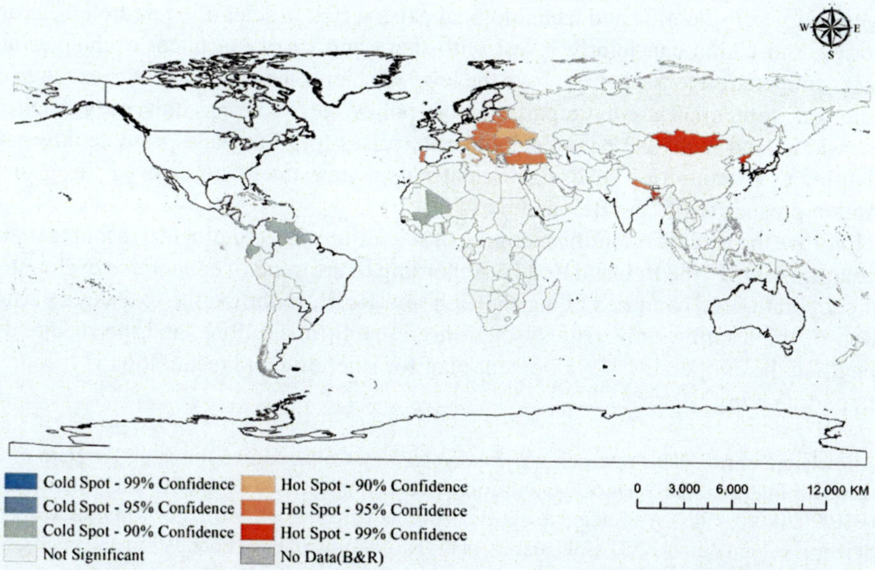

Fig. 1. Spatial pattern of S&T capital resources input based on G-index.

The total number of scientific and technological papers produced from 2005 to 2018 is mainly concentrated in China and some countries in Eastern Europe and Western Asia. Among them, China and Italy are particularly prominent, with the total number of scientific and technological papers produced in China exceeding 81,580 million in 2018, and the total number of scientific and technological papers produced in Italy exceeding 48,000 million in 2018.

Total number of patent applications. From 2005 to 2018, the number of patent applications from countries along the Belt and Road increased rapidly. From 2011, the total number of patent applications entered a period of rapid growth, which slowed down significantly after 2016. After the "Belt and Road" was put forward, the scientific and technological output transformation alliance has been established among the countries along the route, and the scientific and technological output transformation test bases have been built. This has stimulated enterprises along the route to promote the organic combination of industry, academia, and research.

The total number of patent applications from 2005 to 2018 is mainly concentrated in China and some Eastern European and West Asian countries. Among them, China is the most prominent, and the total number of Chinese patent applications has exceeded 162,562 million in 2018.

4 Conclusion

From 2005 to 2018, the output of scientific and technological resources of countries along "the Belt and Road" continued to rise, and it was mainly concentrated in China, Eastern Europe and West Asia. In addition to China, Italy, Turkey and Iran have the largest output of scientific and technological papers, rich in scientific research ideas and theories, and China can jointly invest with them and carry out talent exchanges and project cooperation to further improve the level of scientific research between countries. China can implement a counterpart support policy to encourage university teachers, professional and technical personnel and enterprises to go out and promote advanced technologies to countries with insufficient patent investment, to achieve the goal of common prosperity of "the Belt and Road".

Improve the level of scientific planning of scientific and technological resource inputs in countries along "the Belt and Road". According to the trend of economic development and cooperation of countries along the Belt and Road, mobilize the extensive participation of think tanks, universities, scientific research institutions, and the media, and strengthen the cooperative development plan for scientific and technological resources with participating countries.

Acknowledgements. This research was funded by Key Collaborative Research Program of the Alliance of International Science Organizations(Grant number ANSO-CR-KP-2020-02, ANSO-CR-KP-2020-06), Class A strategic leading science and technology project of Chinese Academy of Sciences (Grant number XDA20030203), National Natural Science Foundation of China (Grant number 32161143029, 42071282).

References

1. Zhang, Z., Jiao, B., Li, Y., et al.: A research on the path of innovation-driven development in regions lacking S&T resources. Research Manage. **39**(2), 46–59 (2018)
2. Liu, Z., Chen, J.: Evaluation of the allocation level of science and technology resources and the regional differences. Sci. Technol. Manage. Res. **39**(11), 93–100 (2019)
3. Lin, L., Qing, L., Bing, L.: The efficiency of scientific and technological resources allocation in China's provinces under environmental constraint: spatiotemporal pattern, evolution mechanism and influencing factors. China Sci. Technol. Forum **06**, 125–135+146 (2019)

4. Ling, F., Qi, Y.: Research on current situation of Jiangsu science and technology resource allocation based on clustering and corresponding analysis. Sci. Technol. Prog. Countermeasures **36**(10), 41–48 (2019)
5. Gao, Y.: An empirical test on the impact of government technology resource portfolio strategy on enterprise innovation performance. Stat. Decis. Making **35**(07), 181–184 (2019)
6. Biao, L., Xiaoping, W., Hongmei, J.: Research on resource allocation efficiency of science and technology based on DEA cross-evaluation. Mod. Manage. Sci. **12**, 35–38 (2011)
7. Li, H., Fan, F., Wang, X.: The spatial-temporal pattern analysis of regional science and technology resource allocation differences. World Geogr. Stud. **22**(04), 159–168 (2013)
8. Li, B., Dong, H.: The study on S&T resources allocation ability of research institutes based on the characteristics of speed. China Soft Sci. **01**, 53–62 (2018)
9. Wei, Z., Renjun, Z.: Evaluation index system of regional scientific and technological resource allocation capacity based on BSC. Stat. Decis. Making **03**, 39–41 (2012)
10. Liu, D., Xu, M., Wang, C., Li, X.: Research on the marine science and technology resource allocation in China. Sci. Technol. Prog. Countermeasures **33**(21), 32–39 (2016)
11. Wang, F., Zhao, H.: An empirical study on regional agglomeration capacity of scientific and technological human resources. China Sci. Technol. Forum **02**, 137–140+144 (2006)
12. Mengmeng, W., Chaoqun, M., Zheng, Y.: An empirical study on the influence of innovation resource agglomeration level on innovation performance of high-tech industries. Sci. Technol. Manage. Res. **35**(09), 13–19+26 (2015)
13. Qingbin, G., Kang, L.: Spatial correlation research on the agglomeration ability of regional science and technology innovation resources-taking Hubei province as an example. Hubei Soc. Sci. **05**, 46–53 (2019)
14. Li, J., Hailan, S.: The influence of mobility of R&D personnel on regional innovation performance based on spatial correlation analysis. Chin. J. Manage. **15**(03), 399–409 (2018)
15. Han, J.: Research on regional green innovation efficiency in China. Res. Finan. Iss. **11**, 130–137 (2012)
16. Ren, Y., et al.: Theoretical model and empirical study on green innovation efficiency. Managing the World **07**, 176–177 (2014)
17. Yin, Q., Cheng, Y.: Regional differences and causes of green innovation efficiency in China. Jiangsu Soc. Sci. **02**, 64–69 (2016)
18. Yi, M., Cheng, X.: Spatial-temporal differentiation and influencing factors of green innovation efficiency in cities in the Yangtze River Economic Belt. Urban Issues. **08**, 31–39 (2018)
19. Liang, W., Qin, T., Liang, P.: Classification measurement and comparison of green low-carbon innovation efficiency in China. Arid Land Resour. Environ. **33**(11), 9–16 (2019)
20. Han, Z., Cheng, X.: Measurement and analysis of R&D investment and innovation efficiency in China. Quant. Tech. Econ. Res. **37**(05), 98–117 (2020)

Author Index

I. N. Vladimirov et al. (Eds.): RERSDNA 2022, SPEES, pp. 195–196, 2023.
https://doi.org/10.1007/978-3-031-28978-1